Contents

Homeschool Testing Book: Algebra 2, 9780547625850

Introduction

The Saxon Homeschool Testing Book for Algebra 2 contains Tests, a Testing Schedule, Test Answer Forms, a Test Analysis Form, and Test Solutions. Descriptions of these components are provided below.

About the Tests

The tests are available after every five lessons, beginning after Lesson 10. The tests are designed to provide students with sufficient time to learn and practice each concept before they are assessed. The test design allows students to display the skills they have developed, and it fosters confidence that will benefit students when they encounter comprehensive standardized tests.

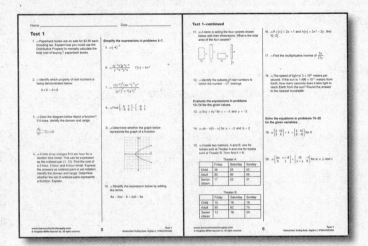
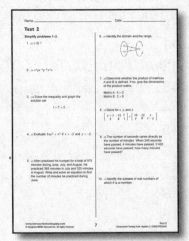

Testing Schedule

Administering the tests according to the schedule is essential. Each test is written to follow a specific five-lesson interval in the textbook. Following the schedule allows students sufficient practice on new topics before they are assessed on those topics.

Tests should be given after every fifth lesson, beginning after Lesson 10. The testing schedule is explained in greater detail on page 4 of this book.

Homeschool Testing Book: Algebra 2, 9780547625850

Optional Test Solution Answer Forms are included in this book. Each form provides a structure for students to show their work.

About the Test Solution Answer Forms

This book contains three kinds of answer forms for the tests that you might find useful. These answer forms provide sufficient space for students to record their work on tests.

Answer Form A: Test Solutions

This is a double-sided master with a grid background and partitions for recording the solutions to twenty problems.

Answer Form B: Test Solutions

This is a double-sided master with a plain, white background and partitions for recording the solutions to twenty problems.

Answer Form C: Test Solutions

This is a single-sided master with partitions for recording the solutions to twenty problems and a separate answer column on the right-hand side.

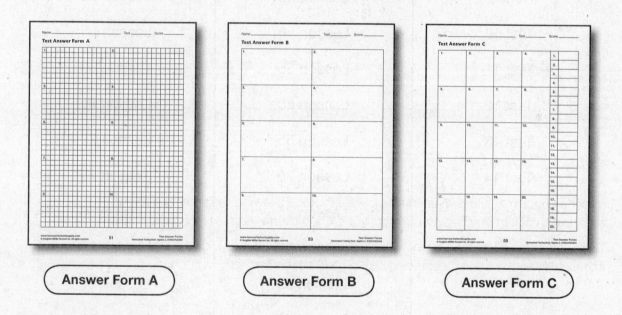

Answer Form A   Answer Form B   Answer Form C

Test Solutions

The Test Solutions are designed to be representative of students' work. Please keep in mind that problems may have more than one correct solution. We have attempted to stay as close as possible to the methods and procedures outlined in the textbook.

3

Testing Schedule

Test to be administered	Covers material through	Give after teaching
Test 1	Lesson 5	Lesson 10
Test 2	Lesson 10	Lesson 15
Test 3	Lesson 15	Lesson 20
Test 4	Lesson 20	Lesson 25
Test 5	Lesson 25	Lesson 30
Test 6	Lesson 30	Lesson 35
Test 7	Lesson 35	Lesson 40
Test 8	Lesson 40	Lesson 45
Test 9	Lesson 45	Lesson 50
Test 10	Lesson 50	Lesson 55
Test 11	Lesson 55	Lesson 60
Test 12	Lesson 60	Lesson 65
Test 13	Lesson 65	Lesson 70
Test 14	Lesson 70	Lesson 75
Test 15	Lesson 75	Lesson 80
Test 16	Lesson 80	Lesson 85
Test 17	Lesson 85	Lesson 90
Test 18	Lesson 90	Lesson 95
Test 19	Lesson 95	Lesson 100
Test 20	Lesson 100	Lesson 105
Test 21	Lesson 105	Lesson 110
Test 22	Lesson 110	Lesson 115
Test 23	Lesson 115	Lesson 120

Homeschool Testing Book: Algebra 2, 9780547625850

Name _____ Date _____

Test 1

1. (1) Paperback books are on sale for $3.95 each including tax. Explain how you could use the Distributive Property to mentally calculate the total cost of buying 7 paperback books.

2. (1) Identify which property of real numbers is being demonstrated below.

$$6 \cdot 9 \doteq 9 \cdot 6$$

3. (4) Does the diagram below depict a function? If it does, identify the domain and range.

4. (4) A bike shop charges $12 per hour for a tandem bike rental. This can be expressed as the ordered pair (1, 12). Find the cost of a 2-hour, 3-hour, and 4-hour rental. Express the answers as ordered pairs in set notation. Identify the domain and range. Determine whether the set of ordered pairs represents a function. Explain.

Simplify the expressions in problems 5–7.

5. (3) $\left(-4\right)^{-2}$

6. (3) $\dfrac{xy^{-2}x^3y^4x^{-6}}{y^{-5}x^{-4}y^3x}$

7. (3) $\dfrac{x\left(x^{-2}\right)^2\left(xy^{-3}\right)^{-2}y^2}{\left(y^2\right)^2x^{-2}\left(y^3\right)^2}$

8. (5) Find $\begin{bmatrix} 6 & 7 \\ -5 & 0 \end{bmatrix} - \begin{bmatrix} -8 & 11 \\ -9 & 15 \end{bmatrix}$.

9. (4) Determine whether the graph below represents the graph of a function.

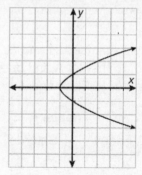

10. (2) Simplify the expression below by adding like terms.

$$6a - 5ba - 8 + 2ab - 8a$$

Homeschool Testing Book: Algebra 2, 9780547625850

Test 1—continued

11. (2) A store is selling the four carpets shown below with their dimensions. What is the total area of the four carpets?

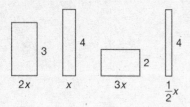

12. (1) Identify the subsets of real numbers to which the number $-\sqrt{7}$ belongs.

Evaluate the expressions in problems 13–14 for the given values.

13. (2) $5xy + 4y^2$ for $x = -3$ and $y = -2$

14. (2) $ab - b(b - a)$ for $a = -3$ and $b = 2$

15. (5) Create two matrices, A and B, one for tickets sold at Theater A and one for tickets sold at Theater B. Then find $A + B$.

Theater A

	Friday	Saturday	Sunday
Child	36	25	42
Adult	80	90	95
Senior citizen	17	22	31

Theater B

	Friday	Saturday	Sunday
Child	15	19	18
Adult	55	62	75
Senior citizen	12	19	25

16. (4) If $c(x) = 2x + 1$ and $h(x) = 3x^2 - 2x$, find $h(-2)$.

17. (1) Find the multiplicative inverse of $\dfrac{3x}{17y}$.

18. (3) The speed of light is 3×10^8 meters per second. If the sun is 1.496×10^{11} meters from Earth, how many seconds does it take light to reach Earth from the sun? Round the answer to the nearest hundredth.

Solve the equations in problems 19–20 for the given variables.

19. (5) $\begin{bmatrix} 2 & 10 \\ 3 & 5 \end{bmatrix} + X = \begin{bmatrix} 3 & 12 \\ 0 & -6 \end{bmatrix}$ for X

20. (5) $\begin{bmatrix} 3w & x + 6 \\ 5 & z \end{bmatrix} = \begin{bmatrix} 15 & -1 \\ x + y & 9 \end{bmatrix}$ for w, x, y, and z.

Homeschool Testing Book: Algebra 2, 9780547625850

Name _____   Date _____

Test 2

Simplify problems 1–2.

1. *(3)* $(-5)^{-2}$

2. *(3)* $x^3yx^{-4}y^{-5}x^2x$

3. *(10)* Solve the inequality and graph the solution set.

$$t + 7 > 5$$

4. *(2)* Evaluate $5xy^2 + x^2$ if $x = -3$ and $y = -2$.

5. *(7)* Allen practiced his trumpet for a total of 975 minutes during June, July, and August. He practiced 365 minutes in July and 325 minutes in August. Write and solve an equation to find the number of minutes he practiced during June.

6. *(4)* Identify the domain and the range.

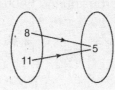

7. *(9)* Determine whether the product of matrices A and B is defined. If so, give the dimensions of the product matrix.

Matrix A: 5×2
Matrix B: 2×6

8. *(5)* Solve for x, y, and z:

$$\begin{bmatrix} y + z & -12 & 7 \\ 3 & 8 & 1 \end{bmatrix} = \begin{bmatrix} 10 & -12 & z \\ y & 8 & x - y \end{bmatrix}$$

9. *(8)* The number of seconds varies directly as the number of minutes. When 240 seconds have passed, 4 minutes have passed. If 420 seconds have passed, how many minutes have passed?

10. *(1)* Identify the subsets of real numbers of which 8 is a member.

Homeschool Testing Book: Algebra 2, 9780547625850

Test 2–continued

11. (1) There are 28 days in February. Rick runs 5 miles each day. Use the Distributive Property to mentally calculate the number of miles Rick runs during February.

12. (6) Change 4.39 to a percent.

13. (4) If $a(x) = 7 + 3x$ and $p(x) = 6x + 2x^2$, find $p(3)$.

14. (10) Solve the inequality.

$$5b < 4 + 3(b + 2)$$

15. (6) The price of a video game is marked up by 35%. What is the new price of the game if the price before the markup was $40?

16. (7) Solve $4n - 8 = 14n + 12$.

17. (9) Find BA if $A = \begin{bmatrix} -4 & 2 & 0 \end{bmatrix}$ and $B = \begin{bmatrix} -5 \\ 3 \\ 6 \end{bmatrix}$.

18. (5) Find the additive inverse matrix of $\begin{bmatrix} 4 & 8 \\ -2 & 4 \\ 9 & -11 \end{bmatrix}$.

19. (2) Sandra needs to buy supplies for a party. Packs of balloons cost $3x$ dollars, packs of cups cost $2x$ dollars, cakes cost $9x$ dollars, and fruit trays cost $7x$ dollars. Sandra buys 3 packs of balloons, 3 packs of cups, 1 cake, and 2 fruit trays. What is the total cost?

20. (8) The cost of a certain type of wire varies directly with its length. The table below shows the cost of the wire at two lengths.

Cost	$15	$20	?
Length	6 feet	8 feet	10 feet

Find the cost of the wire of the third length in the table.

Homeschool Testing Book: Algebra 2, 9780547625850

Name _____ Date _____

Test 3

1. *(12)* Is the equation Work = Force × Distance a joint variation? How do you know?

Simplify problems 2–3.

2. *(3)* $\dfrac{(0.002 \times 10^{-4})(300)}{(0.03 \times 10^{12})(4000 \times 10^{6})}$

3. *(2)* $6xy - 7x - 4yx + 8x + 14$

4. *(4)* Identify the domain and the range.

x	y
−4	9
2	7
8	7

5. *(6)* What is the new amount when 74 is decreased by 40%?

6. *(11)* Write the polynomial $2x - 8 + 3x^2 + 7x^4 - 5x^2$ in standard form. Then identify the leading coefficient and the constant term.

7. *(13)* Graph the equation $y = -2x + 5$ by constructing a table of values.

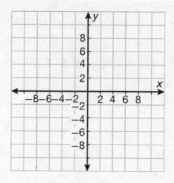

8. *(14)* Evaluate $\begin{vmatrix} 2 & -3 \\ 4 & 0 \end{vmatrix}$.

9. *(10)* Hannah lives 5 miles from school. Ellen lives 9 miles from school. Write an inequality to represent the possible distances between their homes.

10. *(5)* Find $\begin{bmatrix} 2 & 6 \\ -7 & 11 \end{bmatrix} - \begin{bmatrix} -4 & 0 \\ 3 & 8 \end{bmatrix}$.

Homeschool Testing Book: Algebra 2, 9780547625850

Test 3–continued

11. (15) Determine if the linear system below is consistent and independent, consistent and dependent, or inconsistent. If the system is consistent, give the solution.

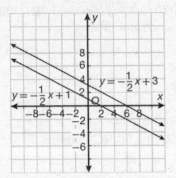

$y = -\frac{1}{2}x + 3$

$y = -\frac{1}{2}x + 1$

12. (1) Identify the subsets of real numbers of which $-\frac{4}{5}$ is a member.

13. (11) Classify $-x^5 + 2x - 3$ by degree and by number of terms.

14. (13) Calculate the slope of the line that contains the following pair of points. Tell whether the line rises, falls, is horizontal, or is vertical.

$$\left(-2,\ 4\right), \left(3,\ -6\right)$$

15. (9) Determine whether the product of matrices A and B is defined. If so, give the dimensions of the matrix.

Matrix A: 3×5, Matrix B: 3×6

16. (2) Evaluate $3ab - a + \dfrac{a - b}{2}$ if $a = -4$ and $b = 6$.

17. (8) The amount of money Jolene gets for yard work is directly proportional to the number of hours she works. If she works for 3 hours, she earns $15. How much does she earn if she does yard work for a total 43 hours throughout the summer?

18. (12) Find the constant of variation for this data set.

x	6	8	10	12
y	10	7.5	6	5

19. (7) Solve $14(r + 1) = 4(r + 4) + 8r$.

20. (14) Find x.

$$\begin{vmatrix} 8 & x + 2 \\ 3 & 4 \end{vmatrix} = 44$$

Homeschool Testing Book: Algebra 2, 9780547625850

Test 4

Name _____ Date _____

1. (3) The density of an object is its mass divided by its volume. The mass of the planet Mercury is 3.3×10^{23} kg. Its volume is 6.1×10^{10} km^3. Approximately what is the density of Mercury in kg/km^3?

2. (5) Find $\begin{bmatrix} -3 & 4 \\ 8 & 0 \\ 2 & 3 \end{bmatrix} - \begin{bmatrix} 7 & 0 \\ 9 & -2 \\ -3 & 8 \end{bmatrix}$.

3. (10) Solve the inequality.

$$6x - 2 \geq 6(x - 2)$$

4. (12) Determine the kind of variation represented by this data. Find the constant of variation and the equation.

x	2	6	8	12
y	13	39	52	78

5. (18) Convert 2400 seconds to hours.

6. (16) Use Cramer's rule to solve.

$$\begin{cases} 2x - 3y = 4 \\ x - y = 3 \end{cases}$$

7. (2) Evaluate $-3b(a - b) + 6ab$ if $a = -4$ and $b = 3$.

8. (8) Distance traveled is directly proportional to the time spent traveling. A car travels 40 kilometers in 30 minutes. At this speed, how far can the car travel in 75 minutes?

9. (14) Evaluate $\begin{vmatrix} -6 & 7 \\ 4 & -3 \end{vmatrix}$.

10. (20) Given $f(x) = 5$; $D = \{$Reals$\}$, $g(x) = 3x$; $D = \{$Reals$\}$, find $(f - g)(x)$ geometrically.

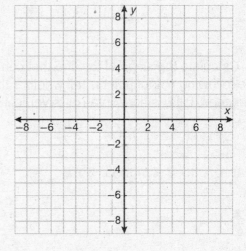

Homeschool Testing Book: Algebra 2, 9780547625850

Test 4–continued

11. *(19)* The length of a sign can be expressed as $(x - 9)$ centimeters. The width of the sign can be expressed as $(x - 7)$ centimeters. Write a polynomial to express the area of the sign.

12. *(1)* Simplify the expression $14 + 9 + 16 + 21$. Identify which property you used for each step.

13. *(17)* Solve $|x - 7| = 2$. Graph the solution.

14. *(15)* Make a table of values for the equations to solve the system.

$3x - y = 8$		$y + 3x = 4$	
x	**y**	**x**	**y**

15. *(13)* Calculate the slope of the line that contains the following pair of points. Tell whether the line rises, falls, is horizontal, or is vertical.

$$(6, -2), (4, -2)$$

16. *(4)* Determine whether the following is a function. Then identify the domain and the range.

17. *(7)* Solve $9n + 25 = 12n - 8$.

18. *(9)* Find AI.

$$A = \begin{bmatrix} -2 & 4 & 3 \\ 3 & -5 & 0 \\ 1 & 6 & 1 \end{bmatrix} \text{ and } I = \begin{bmatrix} 1 & 0 & 0 \\ 0 & 1 & 0 \\ 0 & 0 & 1 \end{bmatrix}$$

19. *(11)* Classify $5x^3 - 14$ by degree and by number of terms.

20. *(6)* Calculate the percent change from 95 to 152 and tell whether the change is a percent increase or percent decrease.

Homeschool Testing Book: Algebra 2, 9780547625850

Name _____   Date _____

Test 5

Solve problems 1–3.

1. (7) $15n - 8 = 7n + 24$

2. (23) $x^2 + 2x - 35 = 0$

3. (24) $\begin{cases} \dfrac{3}{4}x + \dfrac{2}{3}y = 3 \\ \dfrac{x}{2} + 4y = 2 \end{cases}$

4. (15) A video rental store charges nonmembers $5 to rent a video. Members pay $15 per month and $2 for each video they rent. After how many rentals will the total cost be the same for members and nonmembers?

5. (22) Determine whether the graph is continuous, discontinuous, and/or discrete. Determine whether it is a function or a relation. Determine the domain and range of the function.

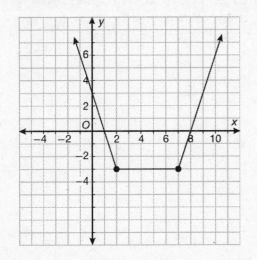

6. (1) Find the multiplicative inverse of $\dfrac{12r}{17s}$.

7. (11) Subtract $(7x^2 - 2x + 5) - (8x^3 - 5x + 14)$.

8. (14) Find the area of the triangle below.

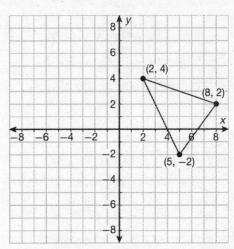

9. (21) This table shows the increase of the populations of the cities of Martin and Taylor from 2005 to 2006. If the populations continue to grow at these constant rates, in about how many years will the two cities have the same population?

City	2005	2006	Increase
Martin	860,417	865,327	4910
Taylor	824,667	836,452	11,785

10. (9) Find AB if $A = \begin{bmatrix} 4 & 5 \\ 1 & -2 \\ 2 & 3 \end{bmatrix}$ and $B = \begin{bmatrix} 3 & 4 \\ 2 & -3 \end{bmatrix}$

 Homeschool Testing Book: Algebra 2, 9780547625850

Test 5–continued

11. *(18)* Simplify using correct significant digits.

$$6.94 + 5.285 + 0.6$$

12. *(17)* Solve $|4x + 24| = 8x$. Check for extraneous solutions.

13. *(20)* Find $(f / g)(4)$ if $f(x) = x + 12$;
$D = \{\text{Integers}\}$; and $g(x) = x + 6$;
$D = \{\text{Positive whole numbers}\}$.

14. *(12)* Find the constant of variation for this data set.

x	2	4	6	8
y	12	6	4	3

15. *(13)* Identify the slope and y-intercept of the line with the equation $y = -\dfrac{1}{4}x + 3$.

Graph the line.

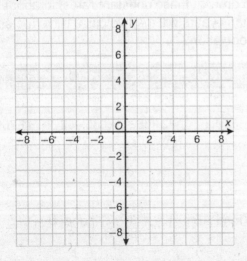

16. *(4)* Identify the domain and range.

x	5	3	8	14
y	12	9	3	9

17. *(19)* Multiply $(4x + 3y)^2$.

18. *(Inv. 1)* Consider the following sentence. "If it is one of the 50 states, and its name begins with H, then it is Hawaii." Write a logic statement for the sentence. Be sure to identify p, q, and r.

19. *(25)* Find the range and standard deviation for the following set of data.

$$2, 4, 6, 8, 10$$

20. *(16)* Use Cramer's rule to solve
$$\begin{cases} 2x + 3y = -4 \\ 6x - 5y = 2 \end{cases}$$

Test 5
Homeschool Testing Book: Algebra 2, 9780547625850

Name _____   Date _____

Test 6

1. (30) Find the equation of the parabola shown.

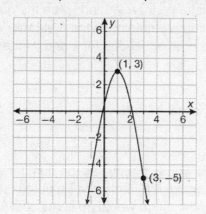

2. (3) Simplify $\dfrac{x^4(x^{-3})^2 y^6 (x^3 y)^{-3}}{(x^3)^3 y (x^{-2})^{-2}}$.

3. (26) A line has slope -3 and passes through the point $(7, 15)$. What is the equation of this line written in slope-intercept form?

4. (11) Add $(5x^3 + x^2 - 3x - 9) + (7 - 3x^2)$.

5. (23) Factor $x^2 - 16x + 64$.

Solve problems 6–7.

6. (7) $-3x + 9 = -18$

7. (24) $\begin{cases} 20 = 5x - 5y \\ x = -8 - 5y \end{cases}$

8. (20) Given $h(x) = x - 9$; $D = \{\text{Reals}\}$, $g(x) = x + 12$; $D = \{\text{Integers}\}$. Find the algebraic sum $(h + g)(x)$.

9. (5) Solve for X: $\begin{bmatrix} 2 & 6 \\ 8 & 3 \end{bmatrix} + X = \begin{bmatrix} 4 & -3 \\ 3 & 7 \end{bmatrix}$.

10. (28) Identify any excluded values. Then simplify the expression.

$$\dfrac{5b^2 - 125}{b - 5}$$

Homeschool Testing Book: Algebra 2, 9780547625850

Test 6—continued

11. (29) Solve the system of equations.

$$\begin{cases} x + y + z = 6 \\ 4x + 2y + z = 9 \\ 7x + 4y + z = 9 \end{cases}$$

12. (2) Evaluate the expression if $a = 2$ and $b = -5$.

$$2ab - b + \frac{a + b}{3}$$

13. (18) Convert 15 yd^2 to in^2.

14. (12) Find the constant of variation for this data set.

x	2	4	6	8
y	24	12	8	6

15. (6) A camera that regularly costs $84 is being offered at a 15% discount. Find the sale price of the camera.

16. (Inv. 2) A car traveling at 55 miles per hour uses 3 gallons of fuel. The fuel tank has a capacity of 25 gallons. Let t be the time in hours, x be the distance traveled, and y be the number of gallons remaining in the fuel tank. Write parametric equations for x and y in terms of t.

17. (8) Cost varies directly as the number of items purchased. If 8 items can be purchased for $44, how much would 36 items cost?

18. (27) Convert $f(x) = 5(x - 3)^2 + 7$ into standard form.

19. (10) Solve the inequality and graph the solution set.

$$-4(3 + b) \leq 12$$

20. (19) Multiply $(4x + 7)(9x - 3)$.

Name _____   Date _____

Test 7

1. (35) Find the zeros of the quadratic function.

    $f(x) = 9x^2 - 49$

2. (9) Find AB if $A = \begin{bmatrix} -2 & -3 \\ -1 & 0 \\ 3 & 5 \end{bmatrix}$ and

    $B = \begin{bmatrix} -2 & -3 \\ 2 & 4 \end{bmatrix}$.

3. (14) Find the determinant of $\begin{bmatrix} -7 & 3 & 2 \\ 5 & -3 & 4 \\ 1 & 0 & 6 \end{bmatrix}$

    using expansion by minors.

4. (28) Identify any excluded values. Then simplify the expression.

    $$\frac{-6x^2 + 18x + 60}{7x^2 + 56x + 84}$$

5. (30) Graph $f(x) = (x - 1)^2 - 3$.

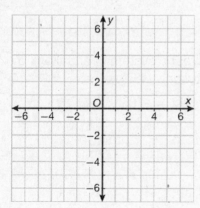

6. (15) Solve this system by graphing.

    $$2x - 3y = 5$$
    $$x + y = 5$$

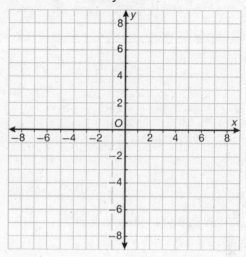

7. (31) Multiply, and then evaluate for $x = 4$:

    $$\frac{9x^2 - 25}{x + 3} \cdot \frac{x^2 + 3x}{3x^2 + 4x - 15}.$$

8. (3) Simplify 4^{-4}.

9. (20) Find $(hg)(-5)$ where $h(x) = x + 6$;

    $D = \{$Reals$\}$, and $g(x) = x - 7$;

    $D = \{$Negative integers$\}$.

10. (29) Solve the system of linear equations. Determine whether the system is consistent or inconsistent.

    $$\begin{cases} 12x + 8y + 18z = -24 \\ -6x - 4y - 9z = 12 \\ 7x + 3y - 8z = 6 \end{cases}$$

Homeschool Testing Book: Algebra 2, 9780547625850

Test 7–continued

11. (34) Graph $3x - 6y = 18$ by solving for y to obtain slope-intercept form.

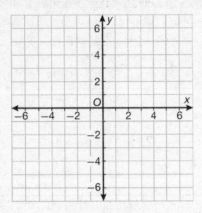

12. (18) Change 624 inches to feet.

13. (32) Find the inverse of $A = \begin{bmatrix} 2 & 4 \\ 1 & -3 \end{bmatrix}$, if it exists.

14. (26) A line with a slope 8 crosses the y-axis at the point (0, 6). What is the equation of this line written in slope-intercept form?

15. (33) There are 5 places to sit at a round table. How many ways can 5 people be seated at the table?

16. (11) Subtract.

$$(7x^3 - 4x^2) - (8x^3 - x^2 + 1)$$

17. (16) Use Cramer's rule to solve.

$$\begin{cases} 4x + 3y = -1 \\ 2x + 5y = 3 \end{cases}$$

18. (22) Graph and state the domain and range of the discrete function $G = \{(-5, 4), (-2, 2), (1,\ 0), (4, -2)\}$.

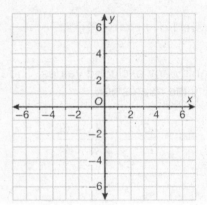

19. (21) Solve the system of equations by substitution.

$$\begin{cases} 4y - 6x = -16 \\ y - 5x = 3 \end{cases}$$

20. (27) Write $9 + 8x = 4x^2 + y$ in standard form.

Homeschool Testing Book: Algebra 2, 9780547625850

Name _____ Date _____

Test 8

1. (13) Calculate the slope of the line that contains the points (4, –3) and (4, –5). Tell whether the line rises, falls, is horizontal, or is vertical.

2. (35) Find the roots of $4x^2 - 5x = 6$.

3. (38) Divide $(x^3 - 9x^2 - 29x + 21)$ by $(x + 3)$.

4. (17) Solve $5|7x| - 10 \le 5$.

5. (39) Graph $3y + 3 > x$ by making a table of values.

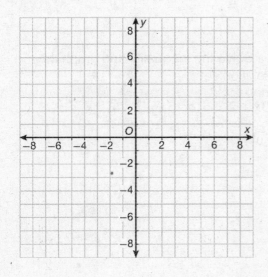

6. (36) Write the equation of the line that is parallel to the graph of $y = -4x + 5$ and crosses the point (3, 6).

7. (24) Solve $\begin{cases} 3x + 2y = -6 \\ -2x + 5y = 23 \end{cases}$ and classify the system.

8. (Inv. 3) Refer to the equation $6x + 4y + 12z = 24$. Find the intercepts. Graph the equation.

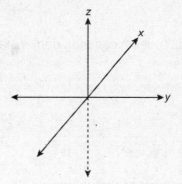

9. (6) Change $\dfrac{14}{5}$ to a percent.

10. (1) Find the additive inverse of $\dfrac{8n}{9}$.

 Homeschool Testing Book: Algebra 2, 9780547625850

Test 8–continued

11. (8) Apples are sold by the bushel or by the peck. The number of bushels varies directly with the number of pecks. In 8 bushels, there are 32 pecks. How many bushels are there in 72 pecks?

12. (33) A student wants to choose a password consisting of six different letters. She places tiles in a bag each with a different letter— A, B, C, D, E, or F—written on each tile. She then randomly chooses a tile from the bag and writes down its letter, without replacing it, until all tiles have been removed. How many passwords are possible using this method?

13. (37) Find the least common denominator.

$$\frac{1}{5x^3} \qquad \frac{1}{6x^5}$$

14. (26) A line passes through points $(-2, -2)$ and $(2, 18)$. What is the equation of the line written in slope-intercept form?

15. (19) Multiply $(4y - 9)^2$.

16. (40) Simplify $\sqrt[3]{-64}$.

17. (29) Solve the system of equations.

$$\begin{cases} x + y + z = 6 \\ 5x + 2y + z = 3 \\ 3x + 3y + z = 6 \end{cases}$$

18. (14) Find the determinant of $\begin{bmatrix} 3 & 2 & -4 \\ 2 & 1 & 5 \\ -2 & 4 & 6 \end{bmatrix}$.

19. (23) A rectangular room is 10 feet long and 15 feet wide. A builder will increase both dimensions by the same amount, resulting in a room that has twice its original floor area. Write and solve an equation to find out by how much the builder should increase each dimension.

20. (10) Solve the inequality and graph the solution set.

$$3x + 2 < 5 \text{ or } 5x - 7 > 8$$

Test 8
Homeschool Testing Book: Algebra 2, 9780547625850

Name _____   Date _____

Test 9

Simplify problems 1–4.
Identify any excluded values.

1. (44) $\dfrac{7}{-3 + \sqrt{3}}$

2. (18) 31.00 feet × 0.020 feet × 12.6 feet

3. (28) $\dfrac{4x - 28}{63 - 9x}$

4. (40) $15\sqrt[3]{27} + 4\sqrt[3]{27} + 7\sqrt{27} - 3\sqrt{27}$

5. (41) Find the value of z. Give the answer in simplest radical form.

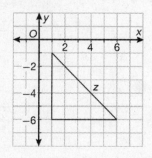

6. (42) Find the permutation of 8 objects taken 3 at a time.

7. (5) Solve for x, y, and z.
$$\begin{bmatrix} 14 & x - y & -3 \\ 7 & z & 3 \end{bmatrix} = \begin{bmatrix} y & -11 & x + z \\ 7 & -6 & 3 \end{bmatrix}$$

8. (38) Divide $14x^3 - 7x^2 + 9x + 4$ by $7x$.

9. (22) Determine whether the graph is continuous, discontinuous, and/or discrete. If the function is discontinuous, name the x-value(s) where the discontinuity occurs.

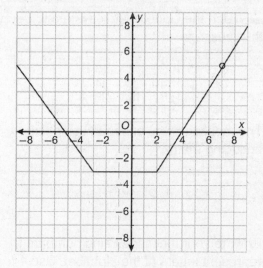

10. (12) The centripetal acceleration, a_c, of an object moving along a circular path is proportional to the square of the object's velocity, v^2, and inversely proportional to the radius, r, of the circle. What kind of variation is this? Write the equation.

Homeschool Testing Book: Algebra 2, 9780547625850

Test 9–continued

11. *(45)* Make a scatter plot of the data set. Sketch a line of best fit, and give the equation of the line.

x	2	4	8	10	14	16	20	22
y	8	6	10	18	20	18	28	20

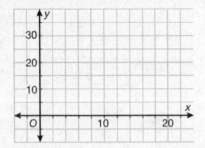

12. *(20)* Given $g(x) = 4x - 9$; $D = \{Reals\}$, $h(x) = x + 8$; $D = \{Integers\}$ find $(g - h)(-3)$ numerically.

13. *(Inv. 4)* Use the cipher-matrix system to decrypt this message.

$$\begin{bmatrix} 35 & 8 & -17 \\ 40 & 9 & -34 \\ 72 & 19 & -29 \end{bmatrix}$$

The message was encoded using this encoding matrix.

$$\begin{bmatrix} 1 & 0 & -1 \\ 2 & 1 & 1 \\ 1 & 0 & -2 \end{bmatrix}$$

Rewrite your message using numbers to represent letters of the alphabet. Use 1 for A, 2 for B, etc.

14. *(37)* Subtract the rational expression.

$$\frac{1}{4 - x} - \frac{3}{4 + x}$$

15. *(16)* Use Cramer's rule to solve.

$$\begin{cases} x + 5y = -4 \\ 3x + 15y = -12 \end{cases}$$

16. *(43)* Determine whether (5, 2), (3, –3), and (2, 7) are solutions of the system of linear inequalities.

$$x \geq -3$$
$$3x + 5y < -5$$

17. *(21)* Solve the system of equations by substitution.

$$\begin{cases} y = x - 5 \\ 3y = 3x - 15 \end{cases}$$

18. *(30)* Determine the vertex and axis of symmetry of $y = -3(x + 2)^2 + 5$.

19. *(31)* Divide, and then evaluate for $x = -2, y = 4$.

$$\frac{46x^6}{9xy^2} \div \frac{23x^5}{9xy^4}$$

20. *(34)* Graph $6x + 2y = 4$ by solving for y to obtain slope-intercept form.

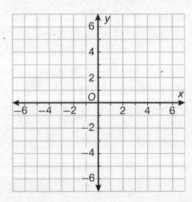

Homeschool Testing Book: Algebra 2, 9780547625850

Name _____   Date _____

Test 10

Simplify problems 1–2.

1. (48) $\dfrac{7 - \dfrac{9}{a}}{\dfrac{4}{b} + 5}$

2. (44) $\dfrac{5 + \sqrt{3}}{7 - 2\sqrt{3}}$

3. (31) Multiply $\dfrac{x + 3}{x^2 - 5x} \cdot (3x^2 - 17x + 10)$.

4. (20) Find $(fg)(-8)$ where $f(x) = x + 2$; $D = \{\text{Reals}\}$, and $g(x) = x - 4$; $D = \{\text{Positive integers}\}$.

5. (43) Ride tickets at a fair cost \$3 each. Tickets to play an arcade game cost \$2 each. Trevor has time for up to a total of 15 rides and games. He can spend up to \$30. Write and graph a system of linear inequalities to represent all the possible combinations of ride and game tickets Trevor can buy.

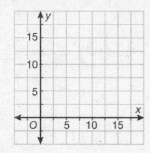

6. (39) Determine whether the point (4, 8) is a solution of the inequality $y > -3x + 7$.

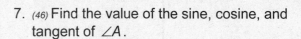

7. (46) Find the value of the sine, cosine, and tangent of $\angle A$.

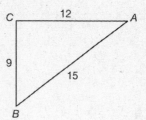

8. (29) Solve the system of equations.

$$\begin{cases} x + 2y + z = -1 \\ 3x + 3y - 2z = -4 \\ x - y + 3z = 12 \end{cases}$$

9. (50) Find an equation for the inverse of $y = 5x - 15$.

10. (7) Solve $9(r - 4) = 8(r + 12) - 5r$.

Homeschool Testing Book: Algebra 2, 9780547625850

Test 10–continued

11. *(47)* Graph $y = \left(\dfrac{1}{3}\right)^x$. Identify the domain, the asymptote, and the range.

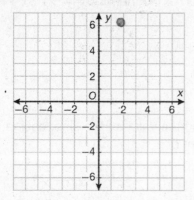

12. *(41)* Find the value of k so that the distance between $(k, 8)$ and $(6, 12)$ is 5.

13. *(24)* Solve $\begin{cases} \dfrac{2}{3}x = 2y \\ 12 + 6y = 4x \end{cases}$.

14. *(36)* What is the relationship between the graphs of $y = -35x + 2$ and $y = -35x - 2$?

15. *(35)* Write a quadratic function that has zeros $-\dfrac{2}{3}$ and 5.

16. *(49)* Use the Binomial Theorem to expand $(a + 4)^5$.

17. *(33)* Pam makes necklaces to sell at craft fairs. The necklaces are either large or small. The stones used in the necklaces are either all red, all blue, all green, or all yellow. How many different types of necklaces does she make?

18. *(11)* Add.

$(9x^3 - 8x^2 + 6x - 12) + (7 - 4x^2 - 2x)$

19. *(25)* Students at a sports meet did the following numbers of pull-ups:

13, 17, 20, 18, 12, 14, 18

Calculate the mean, median, and mode for the number of pull-ups.

20. *(19)* Multiply $(a + 7)(4a^2 - 5a - 3)$.

Homeschool Testing Book: Algebra 2, 9780547625850

Name _____   Date _____

Test 11

1. (52) A surveyor needs to find the distance from point A to point C across a lake. The line AC forms a $45°$ angle with line AB, and the distance from point B to point C is 30 meters, as shown. What is the distance, d, across the lake, rounded to the nearest whole number?

2. (23) Factor $-14x^3 + 35x^2$.

3. (42) Find the combination of 9 objects taken 6 at a time.

4. (6) What is the new amount when 820 is increased by 160%?

5. (31) Divide $\dfrac{x^2 + 4x}{x + 3} \div (8x^4 + 32x^3)$.

6. (51) Use synthetic division to divide $2x^3 - 4x^2 + 3x - 6$ by $x - 3$.

7. (17) Solve $\left|-2x + 7\right| \geq 5$. Graph the solution.

8. (35) Find the roots of $4x^2 + 11x = -6$.

9. (50) Find an equation for the inverse of
$$y = \frac{2}{5}x + 9.$$

10. (54) An artist is preparing jewelry to sell at a craft fair. Rings cost $1.25 to make, and necklaces cost $2.25 to make. The artist will make 60 pieces of jewelry. Usually, about three times as many people buy rings than necklaces. How can the artist keep costs to a minimum while still making sure people have the jewelry they want?

Homeschool Testing Book: Algebra 2, 9780547625850

Test 11—continued

11. (21) Solve the system of equations by substitution.

$$\begin{cases} 3y - 4x = -1 \\ y - 5x = -15 \end{cases}$$

12. (53) Let $f(x) = 3x + 9$ and $g(x) = (x + 4)^2$. Find the composite function $f(g(x))$.

13. (Inv. 5) Determine whether the probability experiment is a binomial experiment. If not, explain why.

Flipping ten coins one time

14. (37) Find any values of x for which the following expression is undefined.

$$\frac{1}{4x^2 - 23x + 15}$$

15. (13) Identify the slope and y-intercept of the line $5x + 3y = 15$. Graph the line.

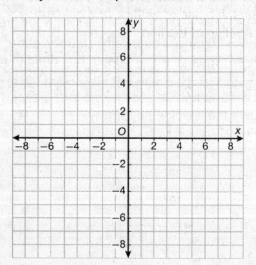

Simplify problems 16–18.

16. (3) $\dfrac{(x^a)^{b+2}(y^a)^{b+5}}{x^{-3a}y^{7a}}$

17. (40) $\sqrt{50h^{15}}$

18. (44) $\dfrac{5}{2\sqrt{11}}$

19. (27) Graph $f(x) = x^2 - 2x + 1$.

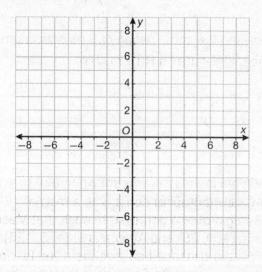

20. (55) A bag contains 20 red and 5 blue marbles. What is the probability of picking out a red marble, keeping it out of the bag, and then picking out a blue marble?

Homeschool Testing Book: Algebra 2, 9780547625850

Name _____   Date _____

Test 12

1. (34) Graph $y - 2 = -\dfrac{3}{4}(x + 4)$.

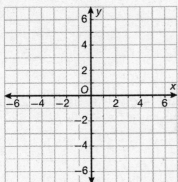

2. (56) Point Q is located on the terminal side of angle θ and has coordinates $(-12, 9)$. Find the exact value of the six trigonometric functions for θ.

3. (16) Use Cramer's Rule to solve

$$\begin{cases} x + 4y = 7 \\ 2x + 2y = 11 \end{cases}$$

4. (41) A cartographer constructs a map of Georgia over a grid system such that Macon is at $(100, 25)$ and Atlanta is at $(175, 125)$. Find the distance between the cities to the nearest kilometer if each unit represents one kilometer.

5. (32) Find the inverse of $B = \begin{bmatrix} -16 & 4 \\ -8 & 2 \end{bmatrix}$, if it exists.

6. (58) Solve by completing the square.

$$x^2 - 16x - 9 = 0$$

7. (60) A jar contains 10 marbles: 4 red, 3 blue, and 3 green. The red marbles are numbered 1–4, the blue are numbered 5–7, and the green are numbered 8–10. Find the probability that a marble drawn from the jar is even or green.

8. (28) Identify any excluded values. Then simplify the expression.

$$\frac{9x^3 - 27x^2}{12x - 36}$$

9. (51) Use synthetic substitution to find $f(-2)$ for $f(x) = x^4 + 3x^2 - 2x + 5$.

10. (45) Make a scatter plot of the data set. Sketch a line of best fit, and give the equation of the line.

x	2	4	8	14	16	22	24	26
y	18	24	16	16	18	14	8	4

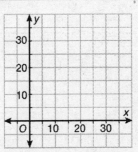

Homeschool Testing Book: Algebra 2, 9780547625850

Test 12–continued

11. (59) Simplify. Assume that all variables are positive.

$$\sqrt[4]{81x^{20}}$$

12. (37) Add the two rational expressions.

$$\frac{1}{9x^4} + \frac{1}{5x^2}$$

13. (46) Find the exact value of the cosecant, secant, and cotangent of $\angle B$.

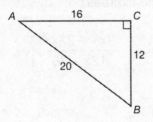

14. (53) Let $f(x) = -4x$, $g(x) = 5x + 3$, and $h(x) = x^2 - 4$. Find the composite function $f(g(h(x)))$.

15. (48) Simplify $\dfrac{\dfrac{2}{x} + \dfrac{1}{x-3}}{\dfrac{5x}{x-3}}$.

16. (9) Find AB if $A = \begin{bmatrix} 1 & 3 \\ -2 & 4 \\ 1 & 2 \end{bmatrix}$ and $B = \begin{bmatrix} 3 & 3 \\ 1 & -2 \end{bmatrix}$.

17. (2) Evaluate $4x^2y + 3y$ if $x = -2$ and $y = -6$.

18. (57) A savings account earns interest at an annual rate of 4%, compounded quarterly. If the account begins with a principal amount of $4000, what will its value be after 3 years?

19. (38) Divide $(x^4 - 12x^2 + 5x + 7)$ by $(x + 3)$.

20. (26) Angela walks at a constant rate of speed from her home to the library. The table shows her distance from the library at different times. These points are on the same line when graphed. Write the equation of this line in slope-intercept form.

Time (minute)	Distance from the Library (meters)
15	5500
20	4500
25	3500

Homeschool Testing Book: Algebra 2, 9780547625850

Test 13

1. *(55)* What is the probability that a point chosen randomly inside the figure is in the shaded region?

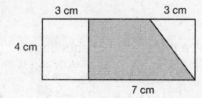

2. *(49)* Use the Binomial Theorem to expand $(2r - 3)^5$.

3. *(27)* Identify the domain and range of $f(x) = -x^2 - 3.8$.

4. *(65)* Solve the equation.

$$x^2 - 4x = 7$$

5. *(14)* Find the determinant of $\begin{bmatrix} -2 & 1 & 4 \\ 6 & 3 & -4 \\ 1 & -1 & 5 \end{bmatrix}$.

6. *(8)* Cost varies directly as the number purchased. If 8 items can be purchased for $28, how much would 44 items cost?

7. *(52)* Use a trigonometric ratio to find the length of s.

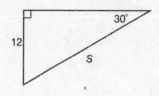

8. *(61)* Factor the expression.

$$x^3 + 6x^2 + 5x + 30$$

9. *(33)* A number is chosen at random from the whole numbers 11 through 30. How many outcomes are in the event *choose a perfect square or a multiple of 7 or a prime number*?

10. *(19)* Multiply $(5x + 7y)(5x - 7y)$.

Homeschool Testing Book: Algebra 2, 9780547625850

11. *(Inv. 6)* Draw a model of algebra tiles to model the following quadratic equation.

$$x^2 + 6x + 9 = 0$$

16. *(63)* Use the unit circle to find the exact value of cos 135°.

12. *(18)* A swimmer's speed is measured to be 4.3 feet per second. What is this speed in miles per hour?

17. *(29)* Solve the system of equations.

$$\begin{cases} 3x + y + 4z = 2 \\ 2x + y - 3z = 13 \\ 5x + y - z = 18 \end{cases}$$

Simplify problems 13–15.

18. *(62)* Solve $-7x^2 = 175$. Write the solutions in terms of i.

13. *(59)* $\left(5^{24}\right)^{\frac{1}{6}}$

19. *(57)* A mass of thallium-201 atoms decays in such a way that half decay about every 3 days. What portion of the original mass remains after 15 days?

14. *(40)* $\dfrac{5}{\sqrt{5} - \sqrt{3}}$

15. *(48)* $\dfrac{\dfrac{3}{a} + 4}{5 - \dfrac{2}{b}}$

20. *(64)* Write the exponential equation in logarithmic form.

$$12^0 = 1$$

Homeschool Testing Book: Algebra 2, 9780547625850

Name _____ Date _____

Test 14

Solve problems 1–3.

1. (7) $\dfrac{2}{3}x - \dfrac{2}{5} = -\dfrac{4}{5}x + 4$

2. (70) $\sqrt{x} - 9 = -16$

3. (17) $|2x + 8| - 12 = 8$

4. (54) A cook is preparing food to sell at a high school football game. At most 200 people are expected to buy a hamburger or hot dog. The cost of preparing a hot dog is $0.25. The cost of preparing a hamburger is $0.35. Usually, three times as many people buy hot dogs as hamburgers. How can the cook keep costs to a minimum while still making sure that most people get their food choice?

5. (15) Solve this system by graphing.

$$3y + 2x = 3$$
$$3y - 2x = -9$$

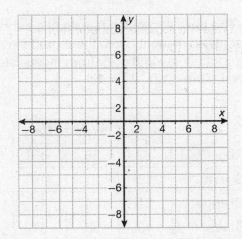

6. (25) Draw a box and whisker plot to display the following data.

$$3, 10, 6, 8, 7, 8, 11, 6, 4$$

7. (61) A rectangular solid with a length of $x - 3$ feet has a volume of $x^3 - 8x^2 - 21x + 108$. Factor the expression for the volume completely.

8. (58) Complete the square and factor the resulting perfect square trinomial.

$$x^2 + 14x$$

9. (32) Solve for matrix X.

$$\begin{bmatrix} 6 & 7 \\ 5 & 6 \end{bmatrix} X = \begin{bmatrix} 3 & -1 \\ 2 & 2 \end{bmatrix}$$

10. (10) Solve and graph the compound inequality.

$$3x + 2 < 8 \text{ or } 4x - 5 > 7$$

Homeschool Testing Book: Algebra 2, 9780547625850

11. (67) Solve $8\sin\theta + 4 = 0$, where θ is any real number of radians.

12. (36) Find the equation of the line that passes through the point $(-10, 4)$ and is perpendicular to the line $y = -\dfrac{2}{5}x - 3$.

13. (22) Determine whether the graph is continuous, discontinuous, and/or discrete. Determine whether it is a function or a relation. Determine the domain and range.

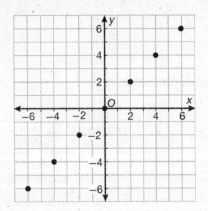

14. (44) Simplify $\sqrt{\dfrac{5}{32}}$.

15. (31) Multiply $\dfrac{x+7}{x^2-4x} \cdot (3x^2 - 7x - 20)$.

16. (63) Find the length of arc s_1. Approximate to the nearest tenth of a centimeter.

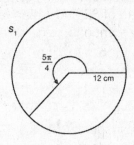

17. (69) Find the absolute value of the number.

$$-9 - 2i$$

18. (42) **ALASKA** has 6 letters with **A** repeated three times. Find the number of distinguishable permutations of letters in **ALASKA**.

19. (66) Identify all the real roots of $x^3 + x^2 - 34x + 56 = 0$.

20. (68) The table shows the results of a survey that asked students from three grade levels to pick their favorite color from those listed. Find $P(\text{Green}\mid 12)$.

	10th	11th	12th
Red	22	38	42
Green	18	21	14
Orange	21	14	11
Blue	43	36	31

Homeschool Testing Book: Algebra 2, 9780547625850

Test 15

Name _____   Date _____

1. (24) Solve and classify the system.

$$\begin{cases} \dfrac{1}{4}x + \dfrac{1}{3}y = 2 \\ x - \dfrac{2}{3}y = 2 \end{cases}$$

2. (Inv. 7) A library gives an evaluation form to each person who checks out a book. People who fill out the evaluation receive a reservation for two to a book talk. Identify the type of sample and determine whether it is biased. Explain your answer.

3. (38) Divide $(20x^3 - 12x^2 + 5x + 2)$ by $4x$.

4. (62) Simplify $-6\sqrt{-64}$.

5. (75) Graph the square root function and its inverse. Determine the domain and range of both functions.

$$y = \sqrt{x - 2}$$

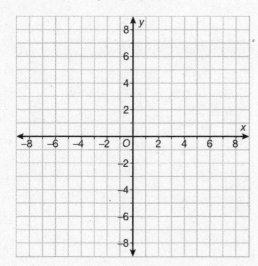

6. (71) Find the area of $\triangle ABC$. Round to the nearest tenth.

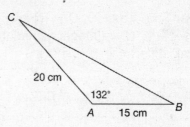

7. (50) Find an equation for the inverse of $y = 2x^2 + 5$. Identify the domain and range of each relation.

8. (58) Solve $x^2 + 18x + 81 = 5$.

9. (74) A farmer has 360 feet of fencing material to enclose a rectangular area of land. He wants the fenced-in area to be 8100 square feet. The equation $w(180 - w) = 8100$ gives the width that meets these requirements. Use the discriminant to explain why these requirements *can* be met.

10. (6) Change $\dfrac{11}{8}$ to a percent.

Test 15–continued

11. (72) Evaluate $\log_4 64$.

12. (20) Given $f(x) = 8x - 9$; $D = \{\text{Reals}\}$, $g(x) = x - 4$; $D = \{\text{Integers}\}$, find the algebraic difference $\{f - g\}(x)$.

13. (33) A student is choosing a five-digit password. Only the digits 1 through 6 are allowed. How many passwords are possible if digits may not be repeated?

14. (11) Classify the polynomial by degree and by number of terms.

$$x^4 + 6x^2 - 8x^4$$

15. (43) Graph the system of linear inequalities.

$$x \geq -6$$
$$x + 4y \leq 8$$

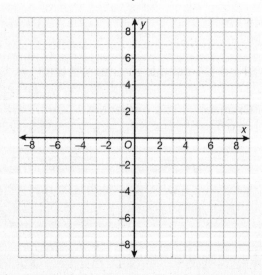

16. (41) The sides of a triangle measure 9 m, 18 m, and 20 m. Is the triangle a right triangle?

17. (66) Find the roots of the equation. Give the multiplicity of each root.

$$x^5 + 14x^4 + 49x^3 = 0$$

18. (73) A researcher visits a park and captures and marks 25 squirrels. On a return visit the next week, the researcher captures 38 squirrels, and 18 of them are marked from the previous week. Estimate the squirrel population in the park.

19. (56) Find the measure of one positive and one negative coterminal angle to the angle $57°$.

20. (67) Solve $6\cos\theta - 3\sqrt{2} = 0$, for $0° \leq \theta \leq 360°$.

Homeschool Testing Book: Algebra 2, 9780547625850

Name _____   Date _____

Test 16

1. (74) Use the discriminant to describe the roots of the equation.

$$9x^2 + 4 = 12x$$

2. (51) Use long division to divide $2x^4 + 2x^3 - x + 6$ by $x^2 - 4x + 2$.

3. (35) Find the roots of the equation. Graph the related function and describe the relationship between the roots, the zeros, and the x-intercepts.

$$x^2 - 4x + 4 = 0$$

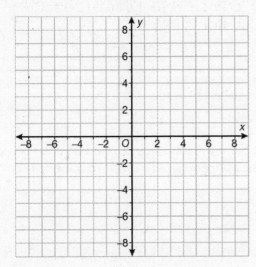

4. (29) Solve the system of equations.

$$\begin{cases} 2x + y + z = 9 \\ -2x + 3y + z = 5 \\ 3x - y + 2z = 10 \end{cases}$$

5. (23) Factor $x^2 - 81$.

Solve problems 6–8.

6. (70) $\sqrt{7x + 4} + 3 = 12$

7. (78) $4x^2 = 324$

8. (59) $x^{\frac{7}{4}} = 128$

9. (55) A calculator is programmed to generate random integers from 1 to 80 inclusive when the enter button is pressed. What is P(multiple of 20) when the enter button is pressed one time?

10. (77) Find b. Round to the nearest tenth.

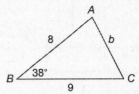

Homeschool Testing Book: Algebra 2, 9780547625850

Test 16–continued

11. *(13)* Calculate the slope of the line that contains the following pairs of points. Tell whether the line rises, falls, is horizontal, or is vertical.

$$(8, 6), (8, 1)$$

12. *(64)* Write the logarithmic equation in exponential form.

$$\log_4 1024 = x$$

13. *(79)* Evaluate the piecewise function for $x = 0$ and $x = 8$.

$$f(x) = \begin{cases} 22, & \text{if } x < 0 \\ -9, & \text{if } 0 \le x < 8 \\ 3, & \text{if } x \ge 8 \end{cases}$$

14. *(46)* Find the value of x. Round to the nearest tenth.

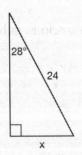

15. *(60)* Two fair coins are flipped two times. Find the probability that both coins come up heads both times.

16. *(80)* The weights of candles produced in a factory have a mean of 9.0 ounces and a standard deviation of 0.25 ounces. A randomly selected candle weighs 9.4 ounces. How many standard deviations is this above the mean?

17. *(37)* Simplify the expression.

$$\frac{x}{x^2 + 2x - 3} - \frac{1}{x^2 + 2x - 3}$$

18. *(76)* Suppose a ball is thrown with an initial speed of $v_o = 7$ from a height $h_o = 25$. To the nearest hundredth of a second, how long will it take the ball to hit the ground? Use the projectile motion equation below.

$$h = -4.9t^2 + v_o t + h_o$$

19. *(9)* Find AB if $A = \begin{bmatrix} 3 & 0 \\ -1 & 5 \\ 2 & 3 \end{bmatrix}$ and $B = \begin{bmatrix} 2 & -1 \\ 0 & 3 \end{bmatrix}$.

20. *(40)* Simplify. $\sqrt{288} - \sqrt{147} - \sqrt{72}$

Homeschool Testing Book: Algebra 2, 9780547625850

Name _____   Date _____

Test 17

1. (Inv. 8) Estimate the area under the curve $y = -x^2 + 3x + 2$ from $0 \le x \le 4$. Use four partitions.

2. (26) A line has slope -4 and passes through the point (9, 12). What is the equation of this line written in slope-intercept form?

3. (78) Solve $4x^3 - 36x = 0$.

4. (85) Find the roots of the polynomial function.

$$y = x^3 + 8$$

5. (39) Graph $3y + 6 > x$ by using slope-intercept form.

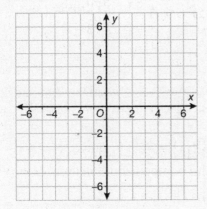

6. (61) Factor the expression.

$$64x^6 + y^6$$

7. (82) Graph the periodic function. Determine the domain and range. Identify the period and amplitude.

$$y = 6\sin(x) + 1$$

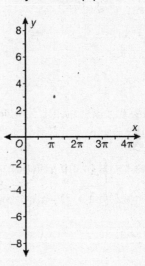

8. (65) Solve the equation.

$$5x^2 + 17x + 6 = 0$$

9. (16) Use Cramer's rule to solve.

$$\begin{cases} 4x + y = 11 \\ -4x + 3y = 1 \end{cases}$$

10. (84) Solve the equation.

$$\frac{3x}{x-4} = 2$$

Test 17–continued

Simplify problems 11–13.

11. *(81)* $\ln e^{4x^2 - 5x}$

12. *(69)* $-5i^{23}$

13. *(3)* $-(-4)^{-3}$

14. *(76)* Find the roots of the polynomial function.

$$f(x) = (x + 5)(x^2 + 7) - (x + 5)(6x + 2)$$

15. *(34)* Graph $x = -5$.

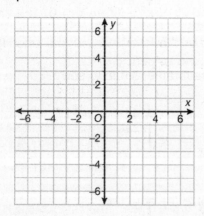

16. *(47)* Suppose $2000 is deposited in a savings account. How much will be in the account in 3 years if the account earns interest at an annual rate of 6%, compounded quarterly?

17. *(83)* Write a quadratic equation whose root is 5.

18. *(31)* Multiply, and then identify all values of x that make the expression undefined.

$$\frac{15x - 5x^2}{x^2 - 4} \cdot \frac{x^2 - 5x + 6}{x^2 - 6x + 9}$$

19. *(21)* Solve the system of equations by substitution.

$$\begin{cases} 3x + 2y = 4 \\ 5x - 3y = 13 \end{cases}$$

20. *(53)* Let $f(x) = 3x + 7$ and $g(x) = 4x$.
Find the composition functions $(f \circ g)(x)$ and $(g \circ f)(x)$.

Homeschool Testing Book: Algebra 2, 9780547625850

Name _____ Date _____

Test 18

1. (86) Graph the periodic function. Determine the domain and range. Identify the period and amplitude.

$$y = 6\cos(x) + 2$$

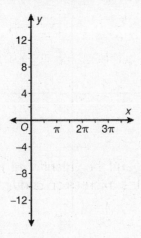

2. (62) Solve the equation. Write the solutions in the form $a + bi$.

$$x^2 + 12x + 38 = 0$$

3. (19) Multiply using the FOIL method.

$$(a + 7)(a - 9)$$

4. (51) Use synthetic division to divide $2x^3 + 2x^2 - 4x - 6$ by $x - 3$.

5. (88) The formula for the area of a trapezoid is $A = \frac{1}{2}(b_1 + b_2)h$. Solve the formula for h.

6. (57) A savings account earns interest at an annual rate of 5%, compounded continuously. If the account begins with a value of $3000, what will its value be after 4 years?

7. (42) Find the number of possible permutations of 5 objects.

8. (87) Use the properties of logarithms to evaluate $\ln(8e)^4$. Round to the nearest tenth.

9. (90) Graph $y = 6\tan(x)$.

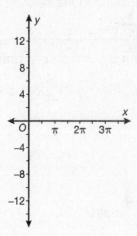

10. (82) Graph the following periodic function and determine its period.

$$y = 6\sin(2x) + 2$$

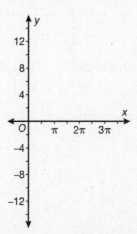

Homeschool Testing Book: Algebra 2, 9780547625850

Test 18–continued

Simplify problems 11–12.

11. *(44)* $\dfrac{4}{\sqrt{5} + \sqrt{3}}$

12. *(48)* $\dfrac{\dfrac{3}{5} + \dfrac{7}{10}}{4 + \dfrac{1}{3}}$

13. *(71)* Find *a*. Round to the nearest tenth.

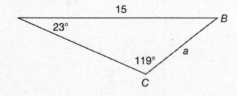

14. *(32)* Solve for matrix *X*.

$$\begin{bmatrix} 2 & 1 \\ 3 & 1 \end{bmatrix} X = \begin{bmatrix} 0 & 6 \\ 9 & -6 \end{bmatrix}$$

15. *(85)* Find the roots of $y = x^3 - 125$.

16. *(75)* Graph the square root function and its inverse. Determine the domain and range of both functions.

$$y = \sqrt{3x - 3}$$

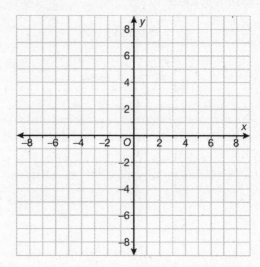

17. *(36)* Find the equation of the line parallel to $y = 6x - 8$ that crosses (5, 2).

18. *(49)* Find the fifth term of the expansion of $(x + 2)^8$.

19. *(89)* Solve $x^2 + 2x > 35$ algebraically.

20. *(5)* Solve for *Y*.

$$Y - \begin{bmatrix} 8 & -2 \\ -3 & 5 \end{bmatrix} = \begin{bmatrix} 6 & 0 \\ -7 & 8 \end{bmatrix}$$

Name _____   Date _____

Test 19

1. *(80)* A set of test scores is normally distributed with a mean of 82 and a standard deviation of 3. What percent of the scores are between 76 and 88?

2. *(Inv. 9)* A landscaper charges $15 per hour of work and rounds the actual work time up to the nearest hour. Write a function to represent the total cost for x hours of work. What type of step function is this function? Explain.

3. *(32)* Find the inverse of the matrix, if it exists.

$$\begin{bmatrix} 4 & -2 \\ -5 & 5 \end{bmatrix}$$

4. *(40)* Simplify the expression.

$$\sqrt{48} \cdot \sqrt{15}$$

5. *(89)* Graph $y \geq -2x^2 + 2x + 4$.

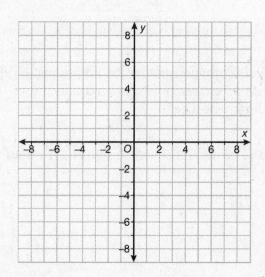

6. *(92)* Find the 18th term of the arithmetic sequence 3, 12, 21, 30, . . .

7. *(77)* Find the area of the triangle. Round to the nearest tenth.

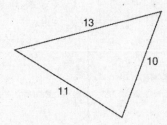

8. *(91)* Write the equation of the circle with center $(-5, 7)$ and radius 6.

9. *(87)* Use the change of base formula to convert $\log_5 (3x)^4$ to base e. Then evaluate when $x = 7$.

10. *(73)* Twenty-four names already numbered are listed below. Five must be randomly chosen for a sample. What five names will be picked for the sample if numbers are chosen from the beginning of the table shown below?

01 Colby	02 Emi	03 Paul	04 Zina
05 Sumi	06 Jim	07 Will	08 Justin
09 Amma	10 Kevin	11 Carl	12 Jody
13 Troy	14 Lee	15 Ben	16 Bob
17 Katy	18 Jhan	19 George	20 Toshi
21 Ahmed	22 Phil	23 Jon	24 Leslie

19872	98729	15372	33792	38273
11199	88927	65928	73911	38075
06978	85463	31228	14775	34293

Homeschool Testing Book: Algebra 2, 9780547625850

Test 19–continued

11. (84) Solve the equation.

$$\frac{30}{x^2 - 9} = \frac{2}{x - 3}$$

12. (79) Write a piecewise function rule for the graph.

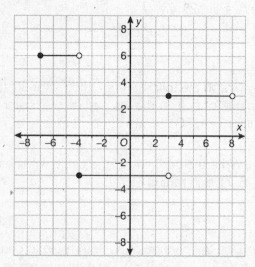

13. (83) The bases of an arch are 60 feet apart. The arch has a height of 40 feet. Write a quadratic function to approximate the arch.

14. (11) Classify $-5x^5 + 3x^2 - 8$ by degree and by number of terms.

15. (94) Solve $\frac{x - 8}{x + 1} > 0$ by finding the sign of the numerator and denominator of the rational expression.

16. (90) Graph $y = \tan(x + \pi)$. Identify its period, undefined values, and phase shift.

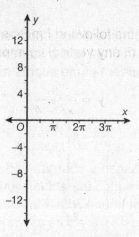

17. (72) Rewrite $\ln\left(\frac{6e}{x}\right)^3$ as a sum or difference of terms.

18. (93) Fluorine-18, with a half-life of 1.83 hours, is used in medical imaging. How long will it take for 5 grams of fluorine-18 to decay to 4 grams?

19. (28) Identify any excluded values. Then simplify the expression.

$$\frac{6x^2 + 2x}{18(3x + 1)^2}$$

20. (95) Determine whether the polynomial $P(x)$ has a zero remainder when divided by $(x - 3)$. Determine $Q(x)$.

$$P(x) = x^5 - 3x^4 - 6x^3 + 20x^2 + 7x - 39$$

Homeschool Testing Book: Algebra 2, 9780547625850

Name _____   Date _____

Test 20

1. (100) Graph the following function. Identify the equations of any vertical asymptotes.

$$y = \frac{4}{x^2 - 4}$$

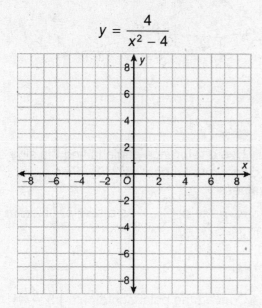

2. (81) Rewrite $\ln\left(\frac{7e^4}{v^8}\right)^9$ as a sum or difference of terms if possible. Then simplify.

3. (70) Solve the equation.

$$-5\sqrt[3]{x - 2} = 15$$

4. (94) Solve $\frac{x - 7}{x + 3} > 0$ by finding the sign of the numerator and the denominator of the rational expression.

5. (37) Find any values of x for which the following expression is undefined.

$$\frac{1}{7x^2 - 65x + 18}$$

6. (99) Find the dot product between the vectors $\begin{bmatrix} -3 \\ 5 \end{bmatrix}$ and $\begin{bmatrix} -2 \\ 9 \end{bmatrix}$.

7. (92) Find a_1 of an arithmetic sequence given that $a_8 = 28$ and $a_{19} = 61$.

8. (98) Graph the following equation. Identify the values of a, b, and c, as well as the major and minor axes. Calculate the eccentricity e.

$$\frac{x^2}{6^2} + \frac{y^2}{3^2} = 1$$

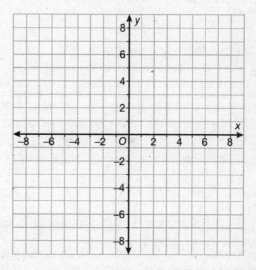

9. (78) Solve $25x^3 - 25x = 0$.

10. (95) Determine whether the polynomial $P(x)$ has a zero remainder when divided by $(x - 3)$. Determine $Q(x)$.

$$P(x) = x^5 - 6x^4 - 3x^3 + 24x^2 + 30x + 18$$

Homeschool Testing Book: Algebra 2, 9780547625850

11. (91) Write the equation of the circle that has a diameter whose endpoints are located at (2, 3) and (6, 5).

12. (68) Suppose the two spinners shown are each spun once. Find the probability that the first spinner lands on a number greater than 2 and the product of the spinners is less than 12.

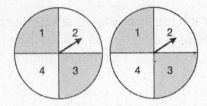

13. (5) Solve for a, b, c, and d.

$$\begin{bmatrix} 21 & b \\ a+c & 12 \end{bmatrix} = \begin{bmatrix} 6+a & -15 \\ 5 & d \end{bmatrix}$$

14. (56) Find the measure of the reference angle for the given angle.

$$\theta = 135°$$

15. (93) If $500 is invested at 7% compounded continuously, how long would it take for the value of the investment to reach $800?

16. (52) Find the length of x given the triangles shown.

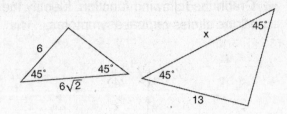

17. (63) Use a reference angle to find the sine, cosine, and tangent of $240°$. Find exact values.

18. (96) Convert the polar coordinate $\left(5, \dfrac{2\pi}{3}\right)$ to Cartesian coordinates.

19. (7) Solve $6n - 8 = 15n + 28$.

20. (97) The fifth term of a geometric sequence is 405. The common ratio is 3. Find the eleventh term.

Name _____   Date _____

Test 21

1. *(Inv. 10)* Write and graph the polar equations for concentric circles centered at the origin with radii 2 and 4.

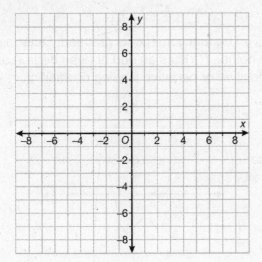

2. *(100)* Factor the terms of the rational function to identify vertical asymptotes.

$$y = \frac{x^2 - 4}{x^2 - 12x + 36}$$

3. *(96)* Convert (–2, 0) to polar coordinates.

4. *(58)* Solve by completing the square.

$$x^2 - 10 + 8 = 0$$

5. *(98)* Graph the following equation. Identify the values of a, b, and c, as well as the major and minor axes. Calculate the eccentricity e.

$$\frac{(x - 1)^2}{5^2} + \frac{(y - 2)^2}{4^2} = 1$$

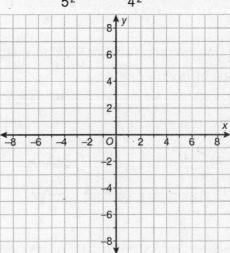

6. *(103)* Graph $y = 3\csc(x)$. Determine its period and asymptotes.

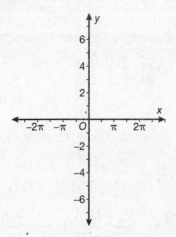

7. *(97)* Find a_8 of a geometric sequence given that $a_3 = 32$ and $a_5 = 512$.

8. *(49)* Use the Binomial Theorem to expand $(f + 3)^5$.

9. *(72)* Simplify $\log_7 7^{9c+2}$.

Homeschool Testing Book: Algebra 2, 9780547625850

Test 21–continued

10. *(105)* An athlete is on a 12-day bicycling plan. On the plan, the athlete is to bike 5 miles on the first day, and on each day thereafter, bike 0.5 mile longer than the previous day. How many total miles will the athlete have biked while on the plan?

11. *(104)* Use the graph of $f(x) = \sqrt{x}$ to sketch the graph of $h(x) = \sqrt{x + 4}$.

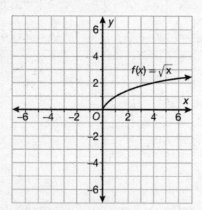

12. *(99)* Find the angle between the vectors
$\begin{bmatrix} -5 \\ 3 \end{bmatrix}$ and $\begin{bmatrix} 2 \\ 4 \end{bmatrix}$.

13. *(31)* Multiply

$$\frac{x^2 - 9}{x + 4} \cdot \frac{4x + 16}{x^2 - 9x + 18} \cdot \frac{x - 6}{x^2 - 4x - 21}.$$

Solve problems 14–15.

14. *(87)* $1000^{8x} = 100$

15. *(102)* $\log x + \log(x + 1) = \log 20$

16. *(92)* Find the 18th term of the arithmetic sequence 12, 5, –2, –9, . . .

17. *(101)* The function $f(x)$ is graphed. Determine whether $f(x)$ has an odd or even degree and a positive or negative leading coefficient.

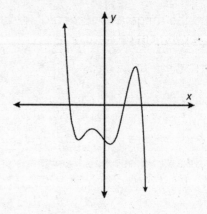

18. *(67)* Evaluate $\cos^{-1}\left(-\dfrac{\sqrt{2}}{2}\right)$ in both radians and degrees if it exists.

19. *(76)* Find the roots of the polynomial function.

$$f(x) = 8x^3 + 10x^2 - 12x$$

20. *(9)* Find AB if $A = \begin{bmatrix} -3 & 2 \\ 2 & 1 \\ 4 & 6 \end{bmatrix}$ and
$B = \begin{bmatrix} 3 & 7 \\ -3 & 2 \end{bmatrix}$.

Homeschool Testing Book: Algebra 2, 9780547625850

Test 22

1. (88) The formula for the period T of a pendulum's swing is $T = 2\pi\sqrt{\dfrac{L}{g}}$. Solve the formula for L.

2. (25) Find the range and standard deviation for the following set of data.

   7, 9, 15, 6, 18

3. (83) Write a quadratic equation whose roots are $6 + 4i$ and $6 - 4i$.

4. (107) Find the equation of the slant asymptote for the following rational function.

   $$f(x) = \frac{x^3 - 2x^2 - 5x + 16}{-x^2 + 6x - 4}$$

5. (103) Graph $y = 4\sec x$. Determine its period and asymptotes.

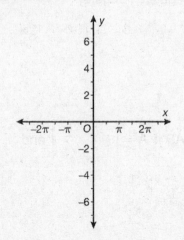

6. (53) Let $f(x) = -4x + 6$, $g(x) = -7x$, and $h(x) = x^2$. Find the composite function $f(g(h(x)))$.

7. (110) Compare the graphs of the three logarithmic functions to see the effect of the constant d on the graph of the function.

   $y = \log(4x)$ $y = \log(4x - 1)$ $y = \log(4x - 3)$

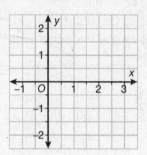

Solve problems 8–10.

8. (102) $\log_7(4x - 17) = 3$

9. (84) $\dfrac{x + 4}{x + 2} + \dfrac{32}{x^2 + 9x + 14} = 3$

10. (78) $(16x^2 - 24x + 9) - (x^2 - 14x + 49) = 0$

Homeschool Testing Book: Algebra 2, 9780547625850

Test 22–continued

11. (61) Factor the expression.

$$2st^3 - 18st - t^2 + 9$$

12. (106) Write the simplest polynomial function with zeros $7 + i$ and $\sqrt{5}$.

13. (109) Graph $\dfrac{x^2}{6^2} - \dfrac{y^2}{3^2} = 1$. Identify the values of a, b, and c, as well as the orientation of the graph. Determine the eccentricity e and the asymptotes.

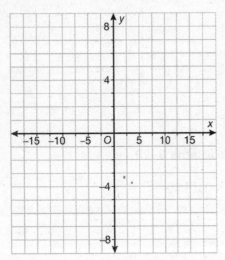

14. (108) Show that $1 + \tan^2(\theta) = \sec^2(\theta)$ for $68.5°$.

15. (101) Identify the leading coefficient and degree and describe the end behavior.

$$R(x) = 4x^3 - 5x^4 + 2x$$

16. (71) Find b. Round to the nearest tenth.

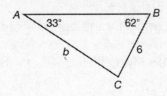

17. (64) Write the logarithmic equation in exponential form.

$$\log_8 512 = x$$

18. (105) Find the sum: $\displaystyle\sum_{k=1}^{32}(4 + 2k)$.

19. (69) Divide $6 + i$ by $9i$. Write the answer in the form $a + bi$.

20. (20) Find $(hg)(-4)$ where $h(x) = 5x + 4$; $D = \{\text{Reals}\}$, and $g(x) = 6x - 3$; $D = \{\text{Negative integers}\}$.

Test 22

Homeschool Testing Book: Algebra 2, 9780547625850

Name _____ Date _____

Test 23

1. (46) A surveyor is located 274.1 feet from the base of the Bank of America Building in Atlanta, GA. The angle of elevation from the surveyor to the top of the building is $75°$. Find the height of the Bank of America Building to the nearest foot.

2. (109) Graph $\dfrac{y^2}{5^2} - \dfrac{x^2}{4^2} = 1$. Identify the values of a, b, and c, as well as the orientation of the graph. Determine the eccentricity e and the asymptotes.

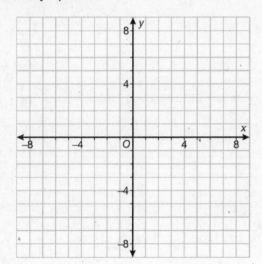

3. (35) Find the roots of $9x = 2x^2 - 35$.

4. (115) Use half-angle identities to find the exact value of $\cos 105°$.

5. (Inv. 11) Use De Moivre's Theorem to evaluate $(3 + 4i)^7$. Express the result in trigonometric form.

6. (86) Graph the function. Determine the domain and range. Identify the period and amplitude.

$$y = 8\cos(x)$$

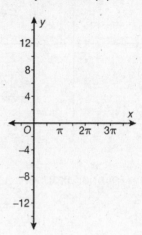

7. (113) Find the sum: $\displaystyle\sum_{k=1}^{7} 5(-2)^{k+2}$.

8. (111) Let $f(x) = 3x^3 + 7x^2 - 5x + 2$. Write a function g that is the reflection of $f(x)$ across the y-axis.

9. (14) Find the determinant of $\begin{bmatrix} -3 & 2 & 0 \\ 4 & -2 & 1 \\ 1 & 5 & 8 \end{bmatrix}$ using expansion by minors.

10. (107) Find the equation of the slant asymptote for the following rational function.

$$f(x) = \dfrac{x^3 - 4x^2 + 3x + 19}{x^2 + 5x - 9}$$

Homeschool Testing Book: Algebra 2, 9780547625850

Test 23—continued

11. *(104)* Use the graph of $f(x) = \sqrt{x}$ to graph
$g(x) = \sqrt{x} - 6$.

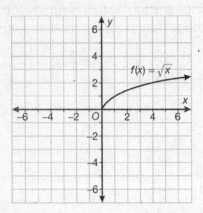

12. *(59)* Write the rational exponent as a radical expression.

$$5^{\frac{1}{7}}$$

13. *(106)* Solve $x^3 - 13x^2 + 56x - 78 = 0$ by finding all the roots.

14. *(114)* Identify the conic section for the standard form equation.

$$\frac{(y+9)^2}{64} + \frac{(x+3)^2}{16} = 1$$

15. *(44)* Simplify $\dfrac{7}{-8 + \sqrt{3}}$.

16. *(74)* A frame maker has 32 inches of framing material with which to make a picture frame. He wants to frame a picture that has an area of 70 square inches. The equation $w(16 - w) = 70$ gives the width of the frame that meets these requirements. Use the discriminant to explain why these requirements can *not* be met.

17. *(21)* Solve the system of equations by substitution.

$$\begin{cases} y = 4x - 6 \\ -5x + 3y = 3 \end{cases}$$

18. *(108)* Show that $\cos^2(\theta) + \sin^2(\theta) = 1$ for $\theta = \dfrac{3\pi}{4}$.

19. *(112)* Find the exact value of $\cos\left(-\dfrac{\pi}{12}\right)$.

20. *(110)* Compare the graphs of the logarithmic functions to see the effect of the constant e on the graph of the function.

$$y = \log(2x) \quad y = \log(2x) + 2 \quad y = \log(2x) + 4$$

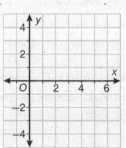

Homeschool Testing Book: Algebra 2, 9780547625850

Name _____   Test _____   Score _____

Test Answer Form A

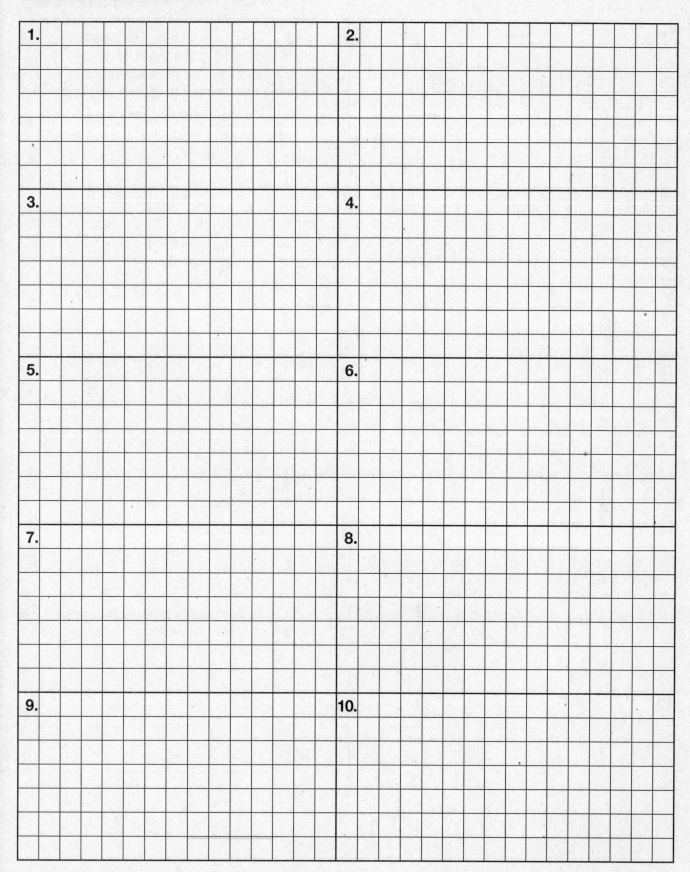

1.

2.

3.

4.

5.

6.

7.

8.

9.

10.

Test Answer Forms
Homeschool Testing Book: Algebra 2, 9780547625850

Name _____ Test _____ Score _____

Test Answer Form A—continued

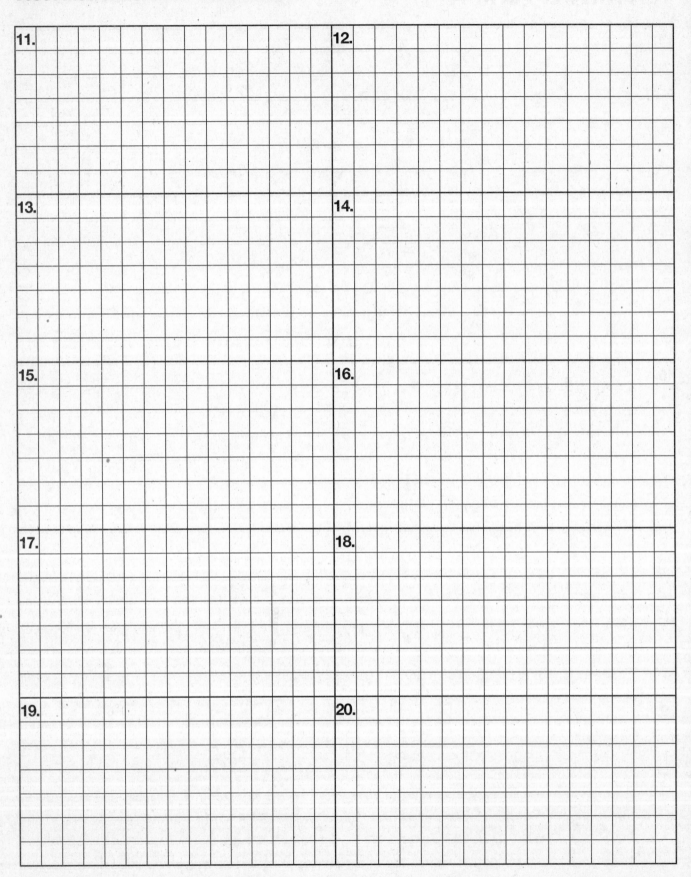

Test Answer Forms

Homeschool Testing Book: Algebra 2, 9780547625850

Name _____ Test _____ Score _____

Test Answer Form B

1.	2.
3.	4.
5.	6.
7.	8.
9.	10.

Homeschool Testing Book: Algebra 2, 9780547625850

Name _____ Test _____ Score _____

Test Answer Form B–continued

11.	12.
13.	14.
15.	16.
17.	18.
19.	20.

Test Answer Forms
Homeschool Testing Book: Algebra 2, 9780547625850

Name _____   Test _____   Score _____

Test Answer Form C

1.	2.	3.	4.	1.	
				2.	
				3.	
				4.	
5.	6.	7.	8.	5.	
				6.	
				7.	
				8.	
9.	10.	11.	12.	9.	
				10.	
				11.	
				12.	
13.	14.	15.	16.	13.	
				14.	
				15.	
				16.	
17.	18	19.	20.	17.	
				18.	
				19.	
				20.	

Test Answer Forms
Homeschool Testing Book: Algebra 2, 9780547625850

Test Analysis Form

Test Item No.	Test Number											
	1	2	3	4	5	6	7	8	9	10	11	12
	Lesson Assessed											
1.	1	3	12	3	7	30	35	13	44	48	52	34
2.	1	3	3	5	23	3	9	35	18	44	23	56
3.	4	10	2	10	24	26	14	38	28	31	42	16
4.	4	2	4	12	15	11	28	17	40	20	6	41
5.	3	7	6	18	22	23	30	39	41	43	31	32
6.	3	4	11	16	1	7	15	36	42	39	51	58
7.	3	9	13	2	11	24	31	24	5	46	17	60
8.	5	5	14	8	14	20	3	Inv. 3	38	29	35	28
9.	4	8	10	14	21	5	20	6	22	50	50	51
10.	2	1	5	20	9	28	29	1	12	7	54	45
11.	2	1	15	19	18	29	34	8	45	47	21	59
12.	1	6	1	1	17	2	18	33	20	41	53	37
13.	2	4	11	17	20	18	32	37	Inv. 4	24	Inv. 5	46
14.	2	10	13	15	12	12	26	26	37	36	37	53
15.	5	6	9	13	13	6	33	19	16	35	13	48
16.	4	7	2	4	4	Inv. 2	11	40	43	49	3	9
17.	1	9	8	7	19	8	16	29	21	33	40	2
18.	3	5	12	9	Inv. 1	27	22	14	30	11	44	57
19.	5	2	7	11	25	10	21	23	31	25	27	38
20.	5	8	14	6	16	19	27	10	34	19	55	26

Test Analysis Form
Homeschool Testing Book: Algebra 2, 9780547625850

Test Analysis Form–continued

Test Item No.	Test Number										
	13	14	15	16	17	18	19	20	21	22	23
	Lesson Assessed										
1.	55	7	24	74	Inv. 8	86	80	100	Inv. 10	88	46
2.	49	70	Inv. 7	51	26	62	Inv. 9	81	100	25	109
3.	27	17	38	35	78	19	32	70	96	83	35
4.	65	54	62	29	85	51	40	94	58	107	115
5.	14	15	75	23	39	88	89	37	98	103	Inv. 11
6.	8	25	71	70	61	57	92	99	103	53	86
7.	52	61	50	78	82	43	77	92	97	110	113
8.	61	58	58	59	65	87	91	98	49	102	111
9.	33	32	74	55	16	90	87	78	72	84	14
10.	19	10	6	77	84	82	73	95	105	78	107
11.	Inv. 6	67	72	13	81	44	84	91	104	61	104
12.	18	36	20	64	69	48	79	68	99	106	59
13.	59	22	33	79	3	71	83	5	31	109	106
14.	40	44	11	46	76	32	11	56	87	108	114
15.	48	31	43	60	34	85	94	93	102	101	44
16.	63	63	41	80	47	75	90	52	92	71	74
17.	29	69	66	37	83	36	72	63	101	64	21
18.	62	42	73	76	31	49	93	96	67	105	108
19.	57	66	56	9	21	89	28	7	76	69	112
20.	64	68	67	40	53	5	95	97	9	20	110

Test Solutions

Test 1

1. Sample: Think of $3.95 as $4.00 − $0.05. Write an expression for the cost of 7 paperback books.

$7(4 - 0.05)$

$7(4) - 7(0.05)$ Use the Distributive Property.

$28 - 0.35$ Multiply.

27.65 Subtract.

The total cost of 7 paperback books is $27.65.

2. Commutative Property of Multiplication

3. Yes, it depicts a function; domain: 7, 9; range: 2

4. {(2, 24), (3, 36), (4, 48)}; domain: 2, 3, 4; range: 24, 36, 48; the set of ordered pairs represents a function since for every value of x there is exactly one y value.

5. $\dfrac{1}{16}$

6. xy^4

7. $\dfrac{1}{x^3 y^2}$

8. $\begin{bmatrix} 14 & -4 \\ 4 & -15 \end{bmatrix}$

9. No, it does not pass the vertical line test.

10. $-3ab - 2a - 8$

11. $18x$

12. $-\sqrt{7}$ is a real number and an irrational number.

13. 46

14. −16

15. $A = \begin{bmatrix} 36 & 25 & 42 \\ 80 & 90 & 95 \\ 17 & 22 & 31 \end{bmatrix}$; $B = \begin{bmatrix} 15 & 19 & 18 \\ 55 & 62 & 75 \\ 12 & 19 & 25 \end{bmatrix}$

$A + B = \begin{bmatrix} 51 & 44 & 60 \\ 135 & 152 & 170 \\ 29 & 41 & 56 \end{bmatrix}$

16. $h(x)$ is 16 when $x = -2$.

17. $\dfrac{17y}{3x}$

18. 498.67 seconds

19. $X = \begin{bmatrix} 1 & 2 \\ -3 & -11 \end{bmatrix}$

20. $w = 5$, $x = -7$, $y = 12$, and $z = 9$

Test 2

1. $\dfrac{1}{25}$

2. $x^2 y^{-4} = \dfrac{x^2}{y^4}$

3. $t > -2$

4. −51

5. $t + 365 + 325 = 975$; $t = 285$ minutes

6. Domain: 8, 11; Range: 5

7. Yes; AB is a 5×6 matrix.

8. $x = 4$, $y = 3$, $z = 7$

9. 7 minutes

10. real numbers, rational numbers, integers, and whole numbers

11. 140 miles

12. 439%

13. 36

14. $b < 5$

15. $54

16. −2

17. $\begin{bmatrix} 20 & -10 & 0 \\ -12 & 6 & 0 \\ -24 & 12 & 0 \end{bmatrix}$

18. $\begin{bmatrix} -4 & -8 \\ 2 & -4 \\ -9 & 11 \end{bmatrix}$

19. $38x$ dollars

20. $25

Homeschool Testing Book: Algebra 2, 9780547625850

Test 3

1. Yes; Sample: work is dependent on two variables, force and distance, and the ratio of work and force × distance is a constant.

2. 5×10^{-25}

3. $2xy + x + 14$

4. Domain: −4, 2, 8; Range: 7, 9

5. 44.4

6. $7x^4 - 2x^2 + 2x - 8; 7; -8$

7.

x	y
−1	7
0	5
1	3

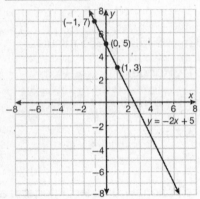

8. 12

9. $4 \leq x \leq 14$

10. $\begin{bmatrix} 6 & 6 \\ -10 & 3 \end{bmatrix}$

11. inconsistent

12. real numbers and rational numbers

13. quintic trinomial

14. −2; falls

15. No; *AB* is undefined.

16. −73

17. $215

18. 60

19. $r = 1$

20. −6

Test 4

1. 5.4×10^{12} kg/km^3

2. $\begin{bmatrix} -10 & 4 \\ -1 & 2 \\ 5 & -5 \end{bmatrix}$

3. $-2 \geq -12$, so the inequality is true for any value of *x*.

4. direct variation; constant of variation is $k = 6.5$; equation is $y = 6.5x$.

5. $\frac{2}{3}$ hr

6. $(5, 2)$

7. −9

8. 100 kilometers

9. −10

10.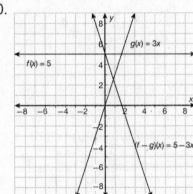

11. $x^2 - 16x + 63$

12. $14 + 16 + 9 + 21$ Commutative Property of Addition
$= (14 + 16) + (9 + 21)$ Associative Property of Addition
$= 30 + 30$ Add.
$= 60$ Add.

13. $x = 5$ or $x = 9$

14. $(2, -2)$
$3x - y = 8$ $y + 3x = 4$

Homeschool Testing Book: Algebra 2, 9780547625850

Sample:

x	y
−1	−11
0	−8
1	−5
2	−2

x	y
−1	7
0	4
1	1
2	−2

15. 0; horizontal

16. function; domain: 9, 14; range: 3

17. $n = 11$

18. $\begin{bmatrix} -2 & 4 & 3 \\ 3 & -5 & 0 \\ 1 & 6 & 1 \end{bmatrix}$

19. cubic binomial

20. 60% increase

Test 5

1. $n = 4$

2. $x = 5, x = -7$

3. $(4, 0)$

4. 5

5. discontinuous function; domain: All real numbers except $x = 2, 7$; range: $y \geq -3$

6. $\dfrac{17s}{12r}$

7. $-8x^3 + 7x^2 + 3x - 9$

8. 15 square units

9. ~4.2 years

10. $\begin{bmatrix} 22 & 1 \\ -1 & 10 \\ 12 & -1 \end{bmatrix}$

11. 12.8

12. $x = 6$

13. $1\dfrac{3}{5}$

14. 24

15.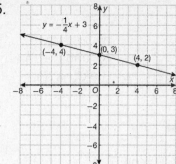

$m = -\dfrac{1}{4}$; y-intercept = 3

16. domain: 3, 5, 8, 14; range: 3, 9, 12

17. $16x^2 + 24xy + 9y^2$

18. Let p = it is one of the 50 states, q = its name begins with H, and r = it is Hawaii. The logic statement is $(p \wedge q) \rightarrow r$.

19. range = 8; standard deviation = $2\sqrt{2}$

20. $x = -\dfrac{1}{2}, y = -1$

Test 6

1. $y = -2(x - 1)^2 + 3$

2. $\dfrac{y^2}{x^{24}}$

3. $y = -3x + 36$

4. $5x^3 - 2x^2 - 3x - 2$

5. $(x - 8)^2$

6. $x = 9$

7. $(2, -2)$

8. $(h + g)(x) = 2x + 3$

9. $\begin{bmatrix} 2 & -9 \\ -5 & 4 \end{bmatrix}$

10. $5(b + 5); b \neq 5$

11. $(2, -3, 7)$

12. −16

13. 19,440 in²

14. 48

15. $71.40

16. $x = 55t; y = 25 - 3t$

Homeschool Testing Book: Algebra 2, 9780547625850

17. $198

18. $5x^2 - 30x + 52$

19. $b \geq -6$

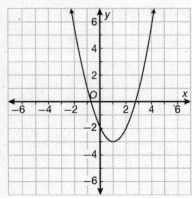

20. $36x^2 + 51x - 21$

Test 7

1. $-\dfrac{7}{3}$ and $\dfrac{7}{3}$

2. $\begin{bmatrix} -2 & -6 \\ 2 & 3 \\ 4 & 11 \end{bmatrix}$

3. 54

4. $-2, -6, \dfrac{-6(x-5)}{7(x+6)}$

5.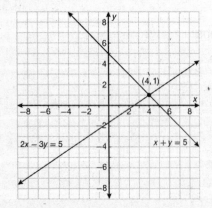

6. $\left(4, 1 \right)$

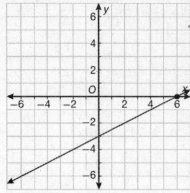

7. $\dfrac{x(3x+5)}{x+3}$; $\dfrac{68}{7}$

8. $\dfrac{1}{256}$

9. -12

10. infinitely many solutions, consistent

11. $y = \dfrac{x}{2} - 3$

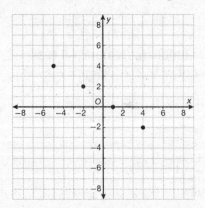

12. 52 ft

13. $\begin{bmatrix} 0.3 & 0.4 \\ 0.1 & -0.2 \end{bmatrix}$

14. $y = 8x + 6$

15. 120

16. $-x^3 - 3x^2 - 1$

17. $\left(-1, 1 \right)$

18. domain: $-5, -2, 1, 4$; range: $-2, 0, 2, 4$

19. $\left(-2, -7 \right)$

20. $y = -4x^2 + 8x + 9$

Test 8

1. undefined; vertical

2. $-\dfrac{3}{4}$ and 2

3. $x^2 - 12x + 7$

4. $-\dfrac{3}{7} \leq x \leq \dfrac{3}{7}$

5.

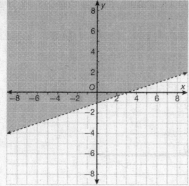

6. $y = -4x + 18$

7. $(-4, 3)$; consistent, independent

8. x: $(4, 0, 0)$; y: $(0, 6, 0)$; z: $(0, 0, 2)$

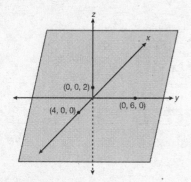

9. 280%

10. $-\dfrac{8n}{9}$

11. 18

12. 720

13. $30x^5$

14. $y = 5x + 8$

15. $16y^2 - 72y + 81$

16. -4

17. $(-1, 1, 6)$

18. -126

19. $(x + 10)(x + 15) = 300$; 5 feet

20. $x < 1$ OR $x > 3$

Test 9

1. $\dfrac{-21 - 7\sqrt{3}}{6}$

2. 7.8 cubic feet

3. $-\dfrac{4}{9}$; $x \neq 7$

4. $57 + 12\sqrt{3}$

5. $5\sqrt{2}$

6. 336

7. $x = 3, y = 14, z = -6$

8. $2x^2 - x + 1 + \dfrac{2x + 4}{7x}$

9. discontinuous function; point of discontinuity at $x = 7$

10. joint variation; $a_c = k\dfrac{v^2}{r}$

11.

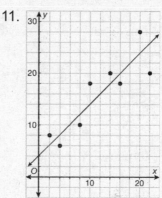

$y = x + 4$

12. -26

13. $\begin{bmatrix} 13 & 8 & 6 \\ 1 & 9 & 21 \\ 20 & 19 & 14 \end{bmatrix}$; MATH IS FUN

14. $-\dfrac{4(x - 2)}{x^2 - 16}$

15. infinite number of solutions

16. $(5, 2)$ no; $(3, -3)$ yes; $(2, 7)$ no

17. $\{(x, y) | y = x - 5\}$

18. vertex: $(-2, 5)$; $x = -2$

19. $2xy^2$; -64

20. $y = -3x + 2$

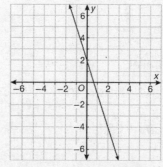

Test 10

1. $\dfrac{7ab - 9b}{4a + 5ab}$

2. $\dfrac{41 + 17\sqrt{3}}{37}$

3. $\dfrac{(x + 3)(3x - 2)}{x}$

4. The common domain is positive integers only, therefore $fg(-8)$ cannot be found.

5.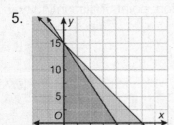

   $x + y \le 15$; $3x + 2y \le 30$

6. yes

7. $\sin A = \dfrac{3}{5}$, $\cos A = \dfrac{4}{5}$, $\tan A = \dfrac{3}{4}$

8. $(3, -3, 2)$

9. $y = \dfrac{1}{5}x + 3$

10. $r = 22$

11.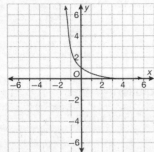

    The domain is the set of all real numbers. The asymptote is the line $y = 0$ (the x-axis). The range is the set of all positive real numbers.

12. 3 and 9

13. $(6, 2)$

14. The lines are parallel.

15. Sample: $f(x) = 3x^2 - 13x - 10$

16. $a^5 + 20a^4 + 160a^3 + 640a^2$
    $+ 1280a + 1024$

17. 8

18. $9x^3 - 12x^2 + 4x - 5$

19. mean: 16; median: 17; mode: 18

20. $4a^3 + 23a^2 - 38a - 21$

Test 11

1. $d \approx 42$ meters

2. $-7x^2(2x - 5)$

3. $C(9, 6) = 84$

4. 2132

5. $\dfrac{1}{8x^2(x + 3)}$

6. $2x^2 + 2x + 9 + \dfrac{21}{x - 3}$

7. $x \ge 6$ or $x \le 1$

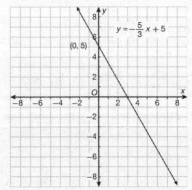

8. $-\dfrac{3}{4}$ and -2

9. $y = \dfrac{5}{2}x - \dfrac{45}{2}$

10. Make no more than 45 rings and no more than 15 necklaces.

11. $(4, 5)$

12. $f\big(g(x)\big) = 3(x + 4)^2 + 9 = 3x^2 + 24x + 57$

13. Not binomial; more than two possible outcomes per trial

14. $x = 5$, $x = \dfrac{3}{4}$

15. $m = -\dfrac{5}{3}$; $(0, 5)$

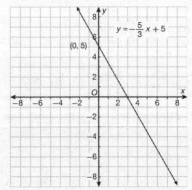

Test Solutions
Homeschool Testing Book: Algebra 2, 9780547625850

16. $x^{ab+5a}y^{ab-2a}$

17. $5h^7\sqrt{2h}$

18. $\dfrac{5\sqrt{11}}{22}$

19.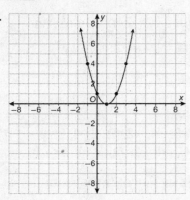

20. $\dfrac{1}{6}$

Test 12

1.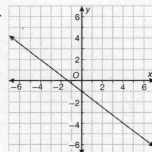

2. $\sin\theta = \dfrac{3}{5}$, $\cos\theta = -\dfrac{4}{5}$, $\tan\theta = -\dfrac{3}{4}$,

   $\csc\theta = \dfrac{5}{3}$, $\sec\theta = -\dfrac{5}{4}$, $\cot\theta = -\dfrac{4}{3}$

3. $\left(5, \dfrac{1}{2}\right)$

4. 125 km

5. The inverse of B does not exist because $|B| = 0$.

6. $x = 8 \pm \sqrt{73}$

7. $\dfrac{3}{5}$

8. 3; $\dfrac{3x^2}{4}$

9. 37

10. Sample: $y = -\dfrac{5}{8}x + 24$

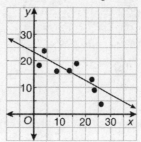

11. $3x^5$

12. $\dfrac{5 + 9x^2}{45x^4}$

13. $\csc B = \dfrac{5}{4}$, $\sec B = \dfrac{5}{3}$, $\cot B = \dfrac{3}{4}$

14. $f(g(h(x))) = -20x^2 + 68$

15. $\dfrac{3x - 6}{5x^2}$

16. $\begin{bmatrix} 6 & -3 \\ -2 & -14 \\ 5 & -1 \end{bmatrix}$

17. -114

18. \$4507.30

19. $x^3 - 3x^2 - 3x + 14 - \dfrac{35}{x + 3}$

20. $y = -200x + 8500$

Test 13

1. $\dfrac{11}{20}$

2. $32r^5 - 240r^4 + 720r^3 - 1080r^2 + 810r - 243$

3. D: all real numbers; R: all real numbers less than or equal to -3.8

4. $2 \pm \sqrt{11}$

5. -92

6. \$154

7. 24

8. $(x + 6)(x^2 + 5)$

9. 11

10. $25x^2 - 49y^2$

Test Solutions
Homeschool Testing Book: Algebra 2, 9780547625850

11.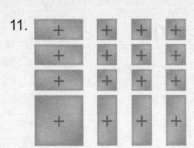

12. 2.9 mph

13. 625

14. $\dfrac{5\sqrt{5} + 5\sqrt{3}}{2}$

15. $\dfrac{3b + 4ab}{5ab - 2a}$

16. $-\dfrac{\sqrt{2}}{2}$

17. $(3, 1, -2)$

18. $\pm 5i$

19. $\dfrac{1}{32}$; 0.03125

20. $\log_{12} 1 = 0$

Test 14

1. $x = 3$

2. no solutions

3. $x = 6$ and $x = -14$

4. Prepare no more than 150 hotdogs and no more than 50 hamburgers.

5. $(3, -1)$

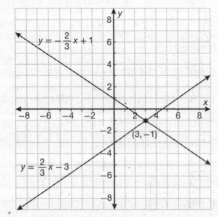

6.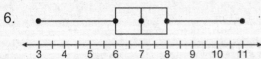

7. $(x - 3)(x - 9)(x + 4)$

8. $(x + 7)^2$

9. $\begin{bmatrix} 4 & -20 \\ -3 & 17 \end{bmatrix}$

10. $x < 2$ or $x > 3$

11. $\dfrac{7\pi}{6} + 2\pi n$ and $\dfrac{11\pi}{6} + 2\pi n$, where n is an integer

12. $y = \dfrac{5}{2}x + 29$

13. Discrete function;
    Domain: $x = -6, -4, -2, 0, 2, 4, 6$;
    Range: $y = -6, -4, -2, 0, 2, 4, 6$

14. $\dfrac{\sqrt{10}}{8}$

15. $\dfrac{(x + 7)(3x + 5)}{x}$

16. $15\pi \approx 47.1$ centimeters

17. $\sqrt{85}$

18. 120

19. $-7, 2, 4$

20. $\dfrac{1}{7} \approx 14.3\%$

Test 15

1. $(4, 3)$, consistent and independent

2. Self-selected sample; biased; the reservation to a book talk is an incentive to fill out the evaluation.

3. $5x^2 - 3x + 1 + \dfrac{x + 2}{4x}$

4. $-48i$

5. For $y = \sqrt{x - 3}$, the domain is $x \geq 3$ and the range is $y \geq 0$. For the inverse, $y = x^2 + 3$, the domain is $x \geq 0$ and the range is $y \geq 3$.

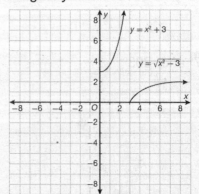

Homeschool Testing Book: Algebra 2, 9780547625850

6. 111.5 cm^2

7. The inverse is $y = \pm\sqrt{\dfrac{1}{2}x - \dfrac{5}{2}}$.

Relation	Domain	Range
$y = 2x^2 + 5$	x is any real number.	$y \geq 5$
$y = \pm\sqrt{\dfrac{1}{2}x - \dfrac{5}{2}}$	$x \geq 5$	y is any real number.

8. $x = -9 \pm \sqrt{5}$

9. Since the discriminant is 0, there is one real solution to the equation $w(180 - w) = 8100$ indicating that there exists one width that meets the requirements.

10. 137.5%

11. 3

12. $7x - 5$

13. 720

14. quartic binomial

15.

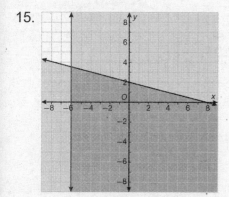

16. no

17. 0 is a root with a multiplicity of 3, and −7 is a root with a multiplicity of 2.

18. about 53 squirrels

19. Sample: 417° and −303°

20. 45° and 315°

Test 16

1. 1 real root

2. $2x^2 + 10x + 36 + \dfrac{123x - 66}{x^2 - 4x + 2}$

3. 2 is the double root of the equation, the only zero of the related function $f(x) = x^2 - 4x + 4$, and the only x-intercept of the graph.

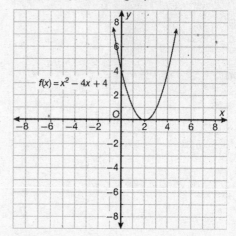

4. (2, 2, 3)

5. $(x + 9)(x - 9)$

6. 11

7. −9 and 9

8. 16

9. 0.05

10. $b \approx 5.6$

11. undefined; vertical

12. $4^x = 1024$

13. $f(0) = -9; f(8) = 3$

14. 11.3

15. 0.0625

16. 1.6

17. $\dfrac{1}{x + 3}$

18. 3.08 seconds

19. $\begin{bmatrix} 6 & -3 \\ -2 & 16 \\ 4 & 7 \end{bmatrix}$

20. $6\sqrt{2} - 7\sqrt{3}$

Test 17

1. 11

2. $y = -4x + 48$

Homeschool Testing Book: Algebra 2, 9780547625850

3. 0, –3, 3

4. $x = -2$

5.

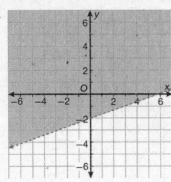

6. $(4x^2 + y^2)(16x^4 - 4x^2y^2 + y^4)$

7.

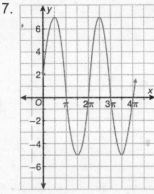

Domain: All real numbers; Range: $-5 \leq x \leq 7$;
Amplitude: 6; Period: 2π

8. $-\dfrac{2}{5}$, –3

9. (2, 3)

10. –8

11. $4x^2 - 5x$

12. $5i$

13. $\dfrac{1}{64}$

14. $x = -5, 5, 1$

15.

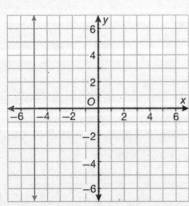

16. $2391.24

17. Sample: $x^2 - 10x + 25 = 0$

18. $\dfrac{-5x}{x + 2}$; 2, –2, 3

19. (2, –1)

20. $(f \circ g)(x) = 12x + 7$; $(g \circ f)(x) = 12x + 28$

Test 18

1.

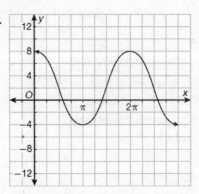

Domain: All real numbers; Range: $-4 \leq x \leq 8$;
Amplitude: 6; Period: 2π

2. $-6 \pm i\sqrt{2}$

3. $a^2 - 2a - 63$

4. $2x^2 + 8x + 20 + \dfrac{54}{x - 3}$

5. $h = \dfrac{2A}{b_1 + b_2}$

6. $3664.21

7. 120

8. 12.3

9.

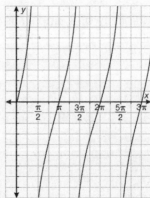

Homeschool Testing Book: Algebra 2, 9780547625850

10.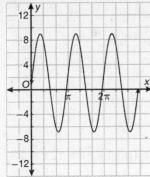

Period: $\dfrac{2\pi}{2} = \pi$

11. $2(\sqrt{5} - \sqrt{3})$

12. $\dfrac{3}{10}$

13. 6.7

14. $X = \begin{bmatrix} 9 & -12 \\ -18 & 30 \end{bmatrix}$

15. $x = 5$

16.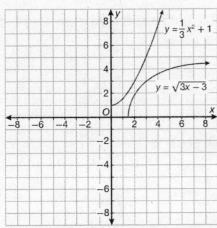

For $y = \sqrt{3x - 3}$, the domain is $x \geq 1$ and the range is $y \geq 0$. For the inverse, $y = \dfrac{1}{3}x^2 + 1$, the domain is $x \geq 0$ and the range is $y \geq 1$.

17. $y = 6x - 28$

18. $1120x^4$

19. $x < -7$ or $x > 5$

20. $Y = \begin{bmatrix} 14 & -2 \\ -10 & 13 \end{bmatrix}$

Test 19

1. 95%

2. $f(x) = 15\lceil x \rceil$; the step function is a least integer function because it rounds up to the nearest integer.

3. $\begin{bmatrix} 0.5 & 0.2 \\ 0.5 & 0.4 \end{bmatrix}$

4. $12\sqrt{5}$

5.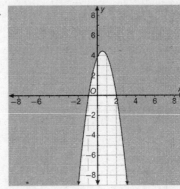

6. 156

7. 53.4 square units

8. $(x + 5)^2 + (y - 7)^2 = 36$

9. 7.5668

10. George, Ben, Jon, Carl, Troy

11. 12

12. $f(x) = \begin{cases} -5 & \text{if } -7 \leq x < -2 \\ 1 & \text{if } -2 \leq x < 4 \\ 6 & \text{if } 4 \leq x < 8 \end{cases}$

13. $y = -0.044x^2 + 2.67x$

14. quintic trinomial

15. $x < -1$ or $x > 8$

16.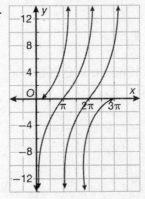

Period: π; Undefined values: $\dfrac{\pi}{2} + n\pi$, where n is an integer; Phase shift: $-\pi$

Homeschool Testing Book: Algebra 2, 9780547625850

17. $3\ln 6 + 3 - 3\ln x$

18. 0.589 hours

19. The excluded value is $x = -\dfrac{1}{3}$; $\dfrac{x}{9(3x+1)}$

20. yes; $x^4 - 6x^2 + 2x + 13$

Test 20

1.
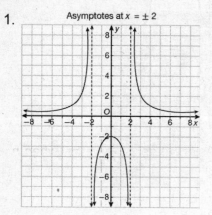

2. $9\ln 7 + 36 - 72\ln v$

3. -25

4. $x < -3$ or $x > 7$

5. $x > 9$, $x = \dfrac{2}{7}$

6. 51

7. 7

8.
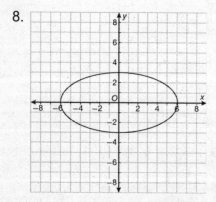

$a = 6$, $b = 3$, $c = 5.1962$, $e = 0.866$,
major $= 12$, minor $= 6$

9. 0, -1, and 1

10. yes; $x^4 - 3x^3 - 12x^2 - 12x - 6$

11. $(x-4)^2 + (y-4)^2 = 5$

12. $\dfrac{5}{16} = 31.25\%$

13. $a = 15$, $b = -15$, $c = -10$, $d = 12$

14. $45°$

15. approximately 6.71 years

16. $13\sqrt{2}$

17. $\sin 240° = -\dfrac{\sqrt{3}}{2}$, $\cos 240° = -\dfrac{1}{2}$,
$\tan 240° = \sqrt{3}$

18. $\left(-\dfrac{5}{2}, \dfrac{5\sqrt{3}}{2}\right)$

19. $n = -4$

20. 295,245

Test 21

1. $r = 2$, $r = 4$;
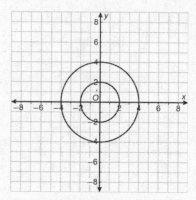

2. Asymptote at $x = 6$

3. $(2, \pi)$

4. $5 \pm \sqrt{17}$

5.

$a = 5$, $b = 4$, $c = 3$, $e = 0.6$, major axis
is 10, minor axis is 8

Test Solutions

Homeschool Testing Book: Algebra 2, 9780547625850

6.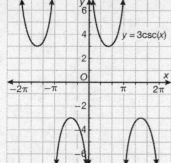

$y = 3\csc(x)$

Period: 2π Asymptotes: $x = \pi n$ when n is any integer

7. 32,768 or −32,768

8. $f^5 + 15f^4 + 90f^3 + 270f^2 + 405f + 243$

9. $9c + 2$

10. 93 miles

11.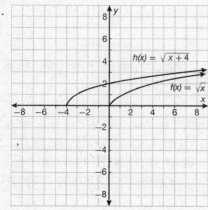

$h(x) = \sqrt{x+4}$

$f(x) = \sqrt{x}$

12. 85.60°

13. $\dfrac{4}{x - 7}$

14. $\dfrac{1}{12}$

15. $x = 4$

16. −107

17. $f(x)$ is of odd degree with a negative leading coefficient.

18. $\dfrac{3\pi}{4}$, or 135°

19. $x = 0, \dfrac{3}{4}, -2$

20. $\begin{bmatrix} -15 & -17 \\ 3 & 16 \\ -6 & 40 \end{bmatrix}$

Test 22

1. $L = g\left(\dfrac{T}{2\pi}\right)^2$

2. Range: 12; Standard deviation: 4.7

3. Sample: $x^2 - 12x + 52 = 0$

4.
5.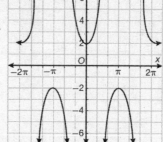

Period: 2π; Asymptotes: $x = \dfrac{\pi}{2} + \pi n$ when n is any integer

6. $f(g(x^2)) = f(-7x^2) = -4(-7x^2) + 6$
$\qquad\qquad\quad = 28x^2 + 6$

7. The horizontal offset is $\dfrac{d}{c}$.

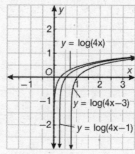

$y = \log(4x)$

$y = \log(4x-3)$

$y = \log(4x-1)$

8. $x = 90$

9. 1 and −9

10. 2 and $-\dfrac{4}{3}$

11. $(2st - 1)(t + 3)(t - 3)$

12. $P(x) = x^4 - 14x^3 + 45x^2 + 70x - 250$

13. Horizontal orientation; $a = 6$; $b = 3$;
$c = 6.7082$; $e = 1.1180$; Asymptotes: $y = \pm\dfrac{1}{2}x$

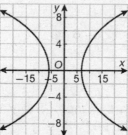

Test Solutions
Homeschool Testing Book: Algebra 2, 9780547625850

14. $1 + \tan^2(68.5°) = 1 + 6.44473 = 7.44473$;
$\sec^2(68.5°) = 1/\cos^2(68.5°) = 7.44473$

15. The leading coefficient is –5. The degree is 4. As $x \to -\infty$, $R(x) \to -\infty$. As $x \to +\infty$, $R(x) \to -\infty$.

16. 9.7

17. $8^x = 512$

18. 1184

19. $\dfrac{1}{9} - \dfrac{2}{3}i$

20. 432

Test 23

1. 1,023 feet

2. Vertical orientation; $a = 5$; $b = 4$; $c = 6.4031$; $e = 1.2806$; Asymptotes: $y = \pm\dfrac{5}{4}$

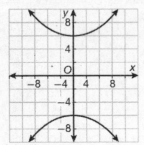

3. $-\dfrac{5}{2}$ and 7

4. $\dfrac{-\sqrt{2} - \sqrt{3}}{2}$

5. $78{,}125\cos(318.78°) + 78{,}125i\sin(318.78°)$

6.

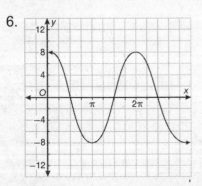

Domain: All real numbers; Range: $-8 \le y \le 8$; Amplitude: 8; Period: 2π

7. –1720

8. $g(x) = -3x^3 + 7x^2 + 5x + 2$

9. 1

10. $f(x) = x - 9$

11.

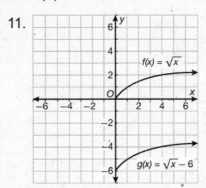

12. $\sqrt[7]{5}$

13. $3,\ 5 + i,\ 5 - i$

14. ellipse

15. $\dfrac{56 + 7\sqrt{3}}{-61}$ or $\dfrac{-56 - 7\sqrt{3}}{61}$

16. The discriminant is –24. Since it is negative, the equation $w(16 - w) = 70$ has no real solutions, so there is no width that can be used with only 32 inches of framing material.

17. $(3, 6)$

18. $\left(-\dfrac{\sqrt{2}}{2}\right)^2 + \left(\dfrac{\sqrt{2}}{2}\right)^2 = \dfrac{1}{2} + \dfrac{1}{2} = 1$

19. $\dfrac{\sqrt{2} + \sqrt{6}}{4}$

20. The graph is vertically translated by the value e.

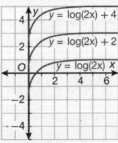

72

Test Solutions

Homeschool Testing Book: Algebra 2, 9780547625850